Surface Engineering of Biomaterials

Editor

Saber AminYavari

MDPI • Basel • Beijing • Wuhan • Barcelona • Belgrade • Manchester • Tokyo • Cluj • Tianjin

Editor
Saber AminYavari
University Medical
Center Utrecht
The Netherlands

Editorial Office
MDPI
St. Alban-Anlage 66
4052 Basel, Switzerland

This is a reprint of articles from the Special Issue published online in the open access journal *Coatings* (ISSN 2079-6412) (available at: https://www.mdpi.com/journal/coatings/special_issues/surf_eng_biomater).

For citation purposes, cite each article independently as indicated on the article page online and as indicated below:

LastName, A.A.; LastName, B.B.; LastName, C.C. Article Title. *Journal Name* **Year**, *Article Number*, Page Range.

ISBN 978-3-03936-898-3 (Hbk)
ISBN 978-3-03936-899-0 (PDF)

Contents

About the Editor

Saber Amin Yavari is an assistant professor at the department of Orthopedics, University Medical Center Utrecht, the Netherlands. He received his Ph.D. degree in Biomechanical Engineering from the Delft University of Technology, the Netherlands, in 2014. His Ph.D. work mainly focused on the development of additive manufacturing technologies to fabricate porous implants. His current research involves the development of multifunctional and translational biomaterials for tissue engineering and advanced drug delivery systems. He has established different surface engineering strategies to prevent implant-associated infections and promote bone regeneration.

Preface to "Surface Engineering of Biomaterials"

Unmet clinical needs, in terms of improved implant fixation, tissue regeneration, infection prevention, and complex reconstructive surgeries, present increasingly more sophisticated challenges that require the development of implants with multiple advanced functionalities. Moreover, a significant increase in life quality and expectancy, together with improvements in surgical techniques, have resulted in a steep increase in implant usage over the past 20 years. Therefore, many attempts have been devoted to the design and synthesis of new biomaterials which could potentially meet these increasing demands. In particular, additive manufacturing or 3D-printing enable us to fabricate biomaterials with much larger surface areas, thereby amplifying the functionalities which originate from their surfaces [1]. The huge surface area of the 3D-printed implants may be treated using various surface biofunctionalization techniques that modify its nano-topography and surface chemistry [2,3]. Furthermore, multifunctional coatings with bespoke release profiles of the active agents provide many opportunities to repair and reconstruct the damaged tissue or organ [4,5]. Nonetheless, there are still many complicated clinical scenarios that should be tackled in this field, and in response, we have focused this Special Issue of Coatings on emerging efforts in the surface engineering of biomaterials and their impact on reducing the abovementioned challenges.

Contributions to this Special Issue include original papers covering the development of different surface modification and coating techniques, which improve the bio-functionality of implants. In particular, Lee et al. [6] immobilized rhBMP-2 and/or rhPDGF-BB on titanium implant surfaces via heparin-dopamine interfaces and evaluated the bone regeneration performance of surface treated alveolar ridges in an animal study (beagle dogs). In an in vitro study, the solution plasma surface modification technique was used by Badruddoza Dithi et al. [7] to apply a thin and uniform hydroxyapatite film coating on titanium implants, which enhanced its bone formation. In another study [8], a novel bioactive glass was coated on a dental implant via electrophoretic deposition, which yielded a firm and non-cytotoxic coating. Furthermore, different surface modification techniques, namely electropolishing, micro-powder blasting and anodizing, were used by Yen et al. [9] to fabricate various micro-nano structure morphologies on an aluminum template. Finally, Muguruma et al. [10] investigated the mechanical and bonding properties of diamond-like carbon coating on stainless steel samples made by the plasma-based ion implantation/ deposition method.

In summary, this Special Issue provides different surface engineering approaches to improve implants' bioactivity, which could potentially be used for (pre-)clinical cases. As such, I hope that this Special Issue will act as a forum to highlight and identify emerging research in the field.

References:

1. S. Amin Yavari *et al.*, Bone regeneration performance of surface-treated porous titanium. *Biomaterials* **35**, 6172-6181 (2014).

2. M. Croes *et al.*, A multifaceted biomimetic interface to improve the longevity of orthopedic implants. *Acta Biomaterialia*, (2020).

3. S. Amin Yavari *et al.*, Antibacterial Behavior of Additively Manufactured Porous Titanium with Nanotubular Surfaces Releasing Silver Ions. *ACS Applied Materials & Interfaces* **8**, 17080-17089 (2016).

4. S. Amin Yavari *et al.*, Layer by layer coating for bio-functionalization of additively manufactured meta-biomaterials. *Additive Manufacturing* **32**, 100991 (2020).
5. F. Jahanmard *et al.*, Bactericidal coating to prevent early and delayed implant-related infections. *Journal of Controlled Release* **326**, 38-52 (2020).
6. S.-H. Lee *et al.*, Effects of Immobilizations of rhBMP-2 and/or rhPDGF-BB on Titanium Implant Surfaces on Osseointegration and Bone Regeneration. *Coatings* **8**, 17 (2018).
7. A. Badruddoza Dithi *et al.*, Application of Solution Plasma Surface Modification Technology to the Formation of Thin Hydroxyapatite Film on Titanium Implants. *Coatings* **9**, 3 (2019).
8. K. Kawaguchi, M. Iijima, K. Endo, I. Mizoguchi, Electrophoretic Deposition as a New Bioactive Glass Coating Process for Orthodontic Stainless Steel. *Coatings* **7**, 199 (2017).
9. M.-L. Yen *et al.*, Aluminum Templates of Different Sizes with Micro-, Nano- and Micro/Nano-Structures for Cell Culture. *Coatings* **7**, 179 (2017).
10. T. Muguruma, M. Iijima, M. Kawaguchi, I. Mizoguchi, Effects of sp2/sp3 Ratio and Hydrogen Content on In Vitro Bending and Frictional Performance of DLC-Coated Orthodontic Stainless Steels. *Coatings* **8**, 199 (2018).

Saber AminYavari
Editor

Article

Effects of Immobilizations of rhBMP-2 and/or rhPDGF-BB on Titanium Implant Surfaces on Osseointegration and Bone Regeneration

So-Hyoun Lee [1,†], Eun-Bin Bae [1,†], Sung-Eun Kim [2,†], Young-Pil Yun [2], Hak-Jun Kim [2], Jae-Won Choi [1], Jin-Ju Lee [1] and Jung-Bo Huh [1,*]

1 Department of Prosthodontics, Dental Research Institute, Institute of Translational Dental Sciences, BK21 PLUS Project, School of Dentistry, Pusan National University, 49 Pusan University-Ro, Yangsan-Si 50612, Gyeongsangnam-Do, Korea; romilove7@hanmail.net (S.-H.L.); 0228dmqls@hanmail.net (E.-B.B.); won9180@hanmail.net (J.-W.C.); ljju1112@hanmail.net (J.-J.L.)

2 Department of Orthopedic Surgery and Rare Diseases Institute, Korea University Medical College, Guro Hospital, #80, Guro-dong, Guro-gu, Seoul 08308, Korea; sekim10@korea.ac.kr (S.-E.K.); ofeel0479@korea.ac.kr (Y.-P.Y.); dakjul@korea.ac.kr (H.-J.K.)

* Correspondence: huhjb@pusan.ac.kr; Tel.: +82-10-8007-9099

† These authors contributed equally to this work.

Received: 23 November 2017; Accepted: 30 December 2017; Published: 31 December 2017

Abstract: The aim of this study was to examine the effects of immobilizing rhPDGF-BB plus rhBMP-2 on heparinized-Ti implants on in vivo osseointegration and vertical bone regeneration at alveolar ridges. Successful immobilizations of rhPDGF-BB and/or rhBMP-2 onto heparinized-Ti (Hepa/Ti) were confirmed by in vitro analysis, and both growth factors were found to be sustained release. To evaluate bone regeneration, rhPDGF-BB, and/or rhBMP-2-immobilized Hepa/Ti implants were inserted into beagle dogs; implant stability quotients (ISQ), bone mineral densities, bone volumes, osseointegration, and bone formation were assessed by micro CT and histometrically. In vivo study showed that the osseointegration and bone formation were greater in the rhPDGF-BB/rhBMP-2-immobilized Hepa/Ti group than in the rhPDGF-BB-immobilized Hepa/Ti group. The rhPDGF-BB/rhBMP-2 immobilized Hepa/Ti group also showed better implant stability and greater bone volume around defect areas and intra-thread bone density (ITBD) than the rhBMP-2-immobilized Hepa/Ti group. However, no significant differences were observed between these two groups. Through these results, we conclude rhBMP-2 immobilized, heparin-grafted implants appear to offer a suitable delivery system that enhances new bone formation in defect areas around implants. However, we failed to observe the synergetic effects for the rhBMP-2 and rhPDGF-BB combination.

Keywords: rhBMP-2; rhPDGF-BB; heparin; implant surface; osseointegration; bone regeneration; beagle dog

1. Introduction

Dental implants have been generally used as credible and secure treatments for the restoration of function and aesthetics of edentulous patients [1]. However, patients who have insufficient bone quality and quantity, or poor healing and regenerative capacities have been reported to experience unfavorable results after implant treatment [2]. To improve the success rate of these patients, it is important to increase the initial fixation of implant fixtures and to shorten the time required for the upper prosthesis to connect [3]. Recently, developments have focused on biomimetic treatment techniques based on applying biomolecules, such as bone morphogenetic protein (BMP) or platelet-derived growth factor (PDGF), to implant surfaces to address these problems [4–6].

BMP is a well-known growth factor that enhances bone regeneration by inducing the differentiation of mesenchymal stem cells to osteoblasts and promotes biosynthesis of bone matrix by regulating factors that are required for osteoinduction [7,8]. BMP-2, which is one of the 16 members of the BMP family, has been proven to be used in a variety of medical treatments by animal and clinical studies [4]. In particular, in one study an anodized titanium implant coated with recombinant human BMP-2 (rhBMP-2) produced by genetic recombination was found to be an effective carrier of rhBMP-2 [5]. However, several studies have reported that rhBMP-2 has no significant effect on bone formation [9,10]. These negative results were suggested to be due to large initial release of rhBMP-2, lack of standardization of the optimal rhBMP-2 concentration, and the use of only one type of growth factor, as natural regeneration process in man involves multiple growth factors [11–14].

Platelet-derived growth factor (PDGF), which is well-characterized tissue growth factor has been used in numerous in vivo and clinical studies [15–20], and has been shown to effectively promote bone, ligament, and cement regeneration in the periodontology field. [21,22]. PDGF is present in bone matrix and is secreted from platelets locally at fracture sites during initial fracture repair [23,24]. PDGF-BB is one of the five PDGF isoforms and is biologically the most potent and binds with greatest affinity to osteoblasts [6,25]. PDGF-BB has both mitogenic and chemotactic effects on osteoblasts and stimulates collagen I synthesis by osteoblasts [23]. It is also important for embryologic skeletal development, and when used topically, it can accelerate fracture healing in animals [26]. PDGF-BB has also been efficaciously used to treat osteoporosis in rodents, in which it improved trabecular bone strength and density [27].

Heparin is a natural linear polysaccharide and a highly sulfated glycosaminoglycan that binds strongly with various growth factors [28]. Biomaterial systems containing heparin exhibit controlled growth factor release [29,30]. When heparin was covalently grafted on anchored free amino positive groups on titanium surfaces, the primary amine groups of growth factors, such as BMP-2 or PDGF-BB, were found to bind to the carboxyl groups of bound heparin [31]. In a previous study, we suggested PDGF-BB/Hepa-Ti system exhibited promising potentials for the enhancements the functions of osteoblast [32]. Also in another previous study on PDGF-BB and BMP-2 co-delivery system on Hep-Ti substrates, the co-delivery system positively promoted functions of osteoblasts [6]. Many experiments have been performed on heparin and growth factor combinations in attempts to induce proper growth factor release, but these experiments were performed at the cellular level or under conditions too far removed from clinical situations.

Therefore, the purpose of this study was to confirm the effects of rhPDGF-BB and rhBMP-2 co-delivery in large animals using clinically reproducible conditions. rhPDGF-BB or/and rhBMP-2 were immobilized onto the surfaces of heparinized-Ti implants and inserted into open defects in beagle dog models. Histomorphometric analysis was conducted to evaluate the effect of rhPDGF-BB, rhBMP-2, and rhPDGF-BB/rhBMP-2 implants on osseointegration and bone regeneration.

2. Materials and Methods

2.1. Materials

Titanium discs (diameter 1.2 cm; height 0.3 cm) were supplied by Cowellmedi (Busan, Korea). Recombinant human platelet-derived growth factor-BB (rhPDGF-BB), recombinant human bone morphogenic protein-2 (rhBMP-2), rhPDGF-BB, and rhBMP-2 ELISA kits were purchased from PeproTech Inc. (Rocky Hill, NJ, USA). Ascorbic acid, dexamethasone, β-glycerophosphate, and dopamine were from Sigma-Aldrich (St. Louis, MO, USA), and heparin sodium (molecular weight: 12,000–15,000 g/moL) was from Acrose Organics (Belgium, NJ, USA). Dulbecco's modified Eagle's medium (DMEM), fetal bovine serum (FBS), phosphate-buffered saline (PBS), and penicillin-streptomycin (PS) were from Gibco BRL (Rockville, MD, USA).

2.2. Surface Modification of Titanium (Ti) with Heparin-Dopamine (Hepa-DOPA) and rhPDGF-BB and/or rhBMP-2

In order to immobilize rhPDGF-BB and/or rhBMP-2, Ti surfaces were modified with heparin-dopamine (Hepa-DOPA). Ti discs were placed in 10 mL Tris·HCl buffer (pH 8.0, 10 mM) containing 2 mg/mL Hepa-DOPA in the darkroom for 24 h. The Hepa-DOPA modified Ti disc was rinsed with distilled water (DW) and dried under nitrogen. Hepa-DOPA modified Ti is hereafter referred to as Heparinized-Ti (Hepa/Ti). To immobilize both rhPDGF-BB and rhBMP-2 on the surface of Hepa/Ti, a Hepa/Ti disc was immersed in MES buffer solution (pH 5.6, 0.1 M), and then rhPDGF-BB (50 ng/mL) and rhBMP-2 (50 ng/mL) were added. The reaction was allowed to proceed for 24 h at room temperature (RT), and then the disc was rinsed with DW and dried. rhPDGF-BB and rhBMP-2 immobilized on Hepa/Ti disc are hereafter referred to as PDGF/BMP/Hepa/Ti disc. rhPDGF-BB (100 ng/mL) or rhBMP-2 (100 ng/mL) modified Hepa/Ti disc were also fabricated using the same method. rhPDGF-BB or rhBMP-2 immobilized on Hepa/Ti disc are hereafter referred to as PDGF/Hepa/Ti or BMP/Hepa/Ti disc, respectively.

2.3. Characterization of Ti, Hepa/Ti, PDGF/Hepa/Ti, BMP/Hepa/Ti, and PDGF/BMP/Hepa/Ti Substrates

2.3.1. Scanning Electron Microscope (SEM) Image

The surfaces of Ti, Hepa/Ti, PDGF/Hepa/Ti, BMP/Hepa/Ti, and PDGF/BMP/Hepa/Ti disc were observed by scanning electron microscopy (SEM; S-2300, Hitachi, Tokyo, Japan). Samples were coated with gold using a sputter coater (Eiko IB, Eiko Engineering, Tokyo, Japan) and SEM was performed at 3 kV.

2.3.2. X-ray Photoelectron Spectroscopy (XPS)

The surface chemical compositions of Ti, Hepa/Ti, PDGF/Hepa/Ti, BMP/Hepa/Ti, and PDGF/BMP/Hepa/Ti disc were investigated by X-ray photoelectron spectroscopy (XPS; K-Alpha spectrometer; Thermo Electron, Rockford, IL, USA). Amounts of heparin immobilized onto Ti were measuredusing toluidine blue. Hepa/Ti disc was immersed in 1 mL PBS buffer (pH 7.4) containing 1 mL 0.005% toluidine blue, gently shaken for 30 min, and 2 mL of hexane was added. After removing the disc, absorbance of the aqueous phase was measured by a Flash Multimode Reader (Varioskan™, Thermo Scientific, Waltham, MA, USA) at 620 nm. The amount of heparin immobilized onto disc were calculated using a calibration curve prepare using different concentrations of heparin.

2.3.3. In Vitro rhPDGF-BB and rhBMP-2 Release

To determine the releases of rhPDGF-BB and rhBMP-2 from PDGF/Hepa/Ti, BMP/Hepa/Ti, and PDGF/BMP/Hepa/Ti, a prepared disc was placed in a 15 mL conical tube (Falcon, North Haledon, NJ, USA) containing PBS buffer (pH 7.4) at 37 °C with 100 rpm. At predetermined times of 1 h, 3, 6, and 10 h, and 1, 3, 5, 7, 10, 14, 21, and 28 days, supernatants were collected and buffer was replenished with an equal volume of fresh PBS. Amounts of rhPDGF-BB and rhBMP-2 released were determined using an enzyme-linked immunosorbent assay kit (ELISA), according to the manufacturer's instruction using a Varioskan Flash Multimode Reader (Thermo Fisher Scientific, Waltham, MA, USA) at 450 nm.

2.4. In Vitro Cell Study

2.4.1. Alkaline Phosphatase (ALP) Activity

To confirm the effects of immobilized rhPDBF-BB, rhBMP-2, or rhPDGF-BB/rhBMP-2 on osteogenic differentiation, we evaluated the ALP activities and calcium contents, as early and late differentiation of MG-63 osteoblast-like cells, were evaluated, respectively. ALP activities were measured after culture for 3, 7, or 10 days. In brief, cells (1×10^5 cells/mL) were seeded on the surfaces of each Ti disc ($n = 5$). At predesignated times, cells and Ti sample were washed with PBS.

Then, RIPA buffer (1×) containing protease and phosphatase inhibitor was added to cells. Cells were then lysed with RIPA (1×) buffer and centrifuged at 13,500 rpm for 1 min to remove cell debris. P-nitrophenyl phosphate solution was then added to supernatants and incubated for 30 min at 37 °C and 1 N NaOH was added to stop reactions. Optical densities of ALP were determined using a Flash Multimode Reader at 405 nm.

2.4.2. Calcium Contents

To determine the calcium contents of MG-63 cells, cells were seeded at a density of 1×10^5 cells/mL on the surfaces of each Ti disc (n = 5) and cultured for 7 or 21 days. Ti discs were then rinsed with PBS, and treated with 0.5 N HCl for 24 h, the Ti discs containing cells were performed by centrifugation at 13,500 rpm for 1 min. Calcium contents were assessed by a QuantiChrom Calcium Assay Kit (DICA-500, BioAssay Systems, Hayward, CA, USA) using calcium chloride as a standard and a Flash Multimode Reader at 612 nm.

2.4.3. Gene Expressions

To assess the osteogenic differentiation effects of different substrates, gene expressions of osteogenic differentiation markers, that is, osteocalcin (OCN) and osteopontin (OPN), were investigated by real-time PCR. Cells were seeded at 1×10^5 cells/mL on Ti, PDGF/Hepa/Ti, BMP/Hepa/Ti, and PDGF/BMP/Hepa/Ti, and cultured for 7 or 21 days (n = 5). RNA was extracted using the RNeasy Plus Mini Kit (Qiagen, Valencia, CA, USA). 1 μg of total RNA was reverse transcribed into cDNA using AccuPower RT PreMix (Bioneer, Daejeon, Korea), according to the manufacturer's protocol. Primer sequences of target genes were as follows: OCN (F) 5′-TTG GTG CAC ACC TAG CAG AC-3′, (R) 5′-ACC TTA TTG CCC TCC TGC TT-3′; and OPN (F) 5′-GAG GGC TTG GTT GTC AGC-3′, (R) 5′-CAA TTC TCA TGG TAG TGA GTT TTC C-3′. PCR amplification and detection were performed using an ABI7300 Real-Time Thermal Cycler (Applied Biosystems, Foster, CA, USA).

2.5. *In Vivo Animal Study*

2.5.1. Fabrication of Implants

Forty implants (Ø 4.0 × H 8.0; Cowellmedi Co., Busan, Korea) were prepared for animal study. All of the the treated implants were fabricated by pure titanium (grade 4), and had microthreads on one end and broader threads on the other. Implant surfaces were anodized (Cowellmedi Co., Busan, Korea), and anodized implants were used as controls (Ti group), the experimental groups were as follows; the heparinized implant group (the Hepa/Ti group), the rhPDGF-BB (0.75 mg/mL) [33] immobilized implant group (the PDGF/Hepa/Ti group), the rhBMP-2 (0.75 mg/mL) immobilized implant group (BMP/Hepa/Ti group), and the rhPDGF-BB (0.75 mg/mL) plus rhBMP-2 (0.75 mg/mL) [34,35] immobilized implant group (PDGF/BMP/Hepa/Ti group). Eight implants were allocated to each group (a total of 40). To apply rhBMP-2/rhPDGF-BB coating, each implant was placed h in the protein solution for 12 (0.75 mg/mL for rhBMP-2, 0.75 mg/mL for rhPDGF-BB) up to its microthreads, and then freeze dried under sterile conditions (freeze dried at −40 °C, and then vacuum dried at ≤20 °C).

This study was carried out with the approval from the Ethics Committee on Animal Experimentation of Chun Nam University (CNU IACUC-TB-2010-10). Five three-year-old beagle dogs of weight 13–15 kg were used for this study. Animals were given two-weeks to acclimatize, fed a soft-dog food diet, and had free access to water.

2.5.2. Surgical Procedures

At first surgery, premolars and first molars of upper and lower jaws were extracted. Animals were pre-anaesthetized with atropine sulfate induction (0.05 mg/kg IM; Dai Han Pharm Co., Seoul, Korea) and maintained on isoflurane (Choongwae Co., Seoul, Korea) gas anesthesia. Lidocaine (1 mL; Yu-Han Co., Gunpo, Korea) containing 1:100,000 epinephrine was infiltrated into mucosae at surgical sites. The upper, lower premolars, and first molars were separated into mesial and distal roots. Care was taken to preserve the lateral, lingual, and buccal walls of alveolar sockets. Teeth were extracted carefully, and extraction sites were sutured with nylon silk (4-0, Mersilk, Ethicon Co., Livingston, UK) to enhance healing. The extraction sites were allowed to heal for two months.

Implant surgery was performed when extraction sockets had completely healed. The anesthesias (local and general) were performed, as described for first surgery. The implants of each groups were implanted at edentulous mandibular alveolar ridge. Briefly, each alveolar ridge was trimmed by ~1.5 mm to create a flat ridge before implant insertion, the buccal open defect model that had 2.5 mm depth was formed. This model was not buccal bone, and there was mesial-lingual-distal 1 mm defect area around 2.5 mm upper portion of implant (Figure 1a). Implants (control (Ti) group, Hepa/Ti group, PDGF/Hepa/Ti group, BMP/Hepa/Ti group, and PDGF/BMP/Hepa/Ti group) were installed randomly on right and left mandibular alveolar ridges (8 implants per dog). To place implants at the same position on both sides, exposed bone was marked at implant placement sites using a ruler. 5 mm of implant was placed within the reduced alveolar ridge to the reference notch level (shown on the implant), which resulted in a 2.5 mm peri-implant buccal open defects (Figure 1b,c). Each implant was covered with cover-screw. Mucoperiosteal flaps were advanced, adapted, and sutured leaving the implants submerged.

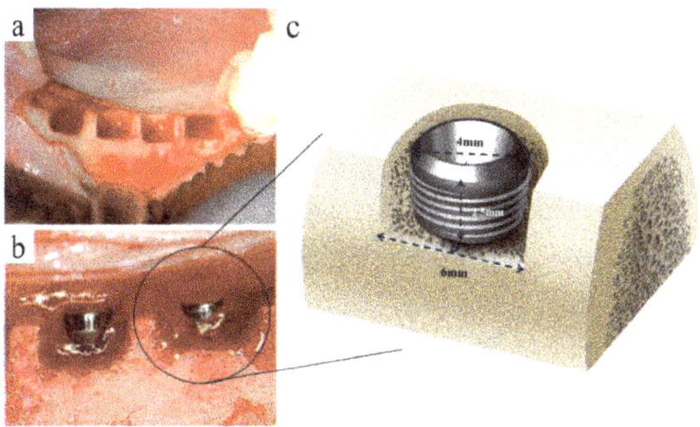

Figure 1. (**a**) Alveolar bone was flattened without exposing cancellous bone; (**b**) 5 mm of implant was placed within the reduced alveolar ridge and upper 2.5 mm of implants was placed in supra alveolar peri-implant buccal open defects; and (**c**) Schematic diagram of the buccal open defect model.

2.5.3. Post-Operative Care after Implant Placement and Sacrifice

A broad spectrum antibiotic (penicillin G with was administered immediately after implant placement and again 48 h later by intramuscular injection (1 mL/5 kg). To control the plaque, Teeth were washed out with 2% chlorhexidine gluconate every day until study completion. Observations of experimental sites with regards to mucosal health, edema, maintenance of suture closure, and tissue infection or necrosis were made daily until suture removal. Suture materials were removed one week after implant placement. Experimantal animals were given a soft diet for two weeks, followed by a conventional regular diet. The animals were anesthetized and euthanized at eight weeks after implant placement by intravenous injection of concentrated sodium pentobarbital (Euthasol, Delmarva Laboratories Inc., Midlothian, VA, USA). Following euthanasia, block sections including implants, alveolar bone, and surrounding mucosa were collected.

2.5.4. Measurment of Implant Stability

Implant stability quotient (ISQ) values were measured to evaluate stability at the time of placement. ISQ values of all the implants placed in mandibles were measured immediately and at week eight after second surgery using Osstell Mentor® (Integtration Diagnostic Ltd., Göteborg, Sweden). ISQ values were measured five times for each implant, and mean and standard deviations (SDs) were calculated after excluding minimum and maximum values.

2.5.5. Micro Computed Tomography (μCT)

The collected samples were fixed in phosphate-buffered formaldehyde (pH 7.4, 0.1 M PBS) and dehydratied in ethanol 70%. Three dimensional (3D) μCT images were obtained to determine bone mineral densities and bone volumes surrounding implants in defect areas. Specimens were wrapped using Parafilm M® (Pechiney Plastic Packaging, Chicago, IL, USA) to prevent dry during scanning, and scanned at 130 kV and 60 μA with a resolution of 12μm pixels using a bromine filter (0.25 mm) (Skyscan-1173 Skyscan®, Kontich, Belgium). In addition, calibration rods of standard bone mineral densities were also scanned. Cone-beam reconstruction (version 2.15, Skyscan®, Kontich, Belgium) was performed, and all scan and reconstruction parameters that were applied were identical for all the specimens and calibration rods. The collected data were analyzed by a CT analyser (version 1.4, Skyscan®, Kontich, Belgium). The region of interest (ROI) was defined as annular region of thickness 1 mm surrounding a defect area in the marginal peri-implant from the first microthread to the last microthread. Bone volumes (mm^3) were measured in this region (Figure 2) and were expressed as percentages of the total ROI volumes (mm^3).

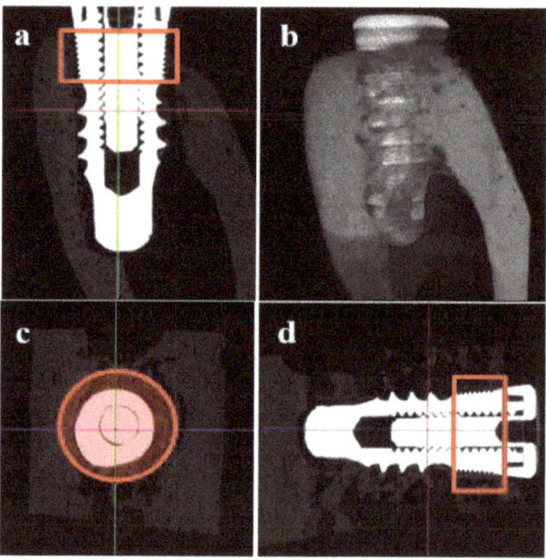

Figure 2. Micro-computed tomography (μCT) images in each group. (**a**) Buccolingual section image; (**b**) three-dimensional (3D) image; (**c**) Horizontal section image; and (**d**) Mesiodistal section image. Region of interest (ROI) was defined as an annular area of thickness 1 mm surrounding the defect area (red circle) in the marginal portion of the peri-implant from the first microthread to the last microthread. Bone volumes were measured in this ROI.

2.5.6. Histologic and Histometric Analysis

The harvested specimens were imsersed in neutral buffered formalin (Sigma Aldrich, St Louis, MO, USA), fixed for two weeks, and dehydrated in ascending concentrations of ethanol (70%, 80%, 90%, and 100%), and embedded in Technovit 7200 VLC resin (Heraeus KULZER, South Bend, IN, USA). Embedded specimen blocks were sectioned longitudinally from the center of implant using an diamond cutter (KULZER EXAKT 300, EXAKT, Norderstedt, Germany). The final slides (30 μm) were prepared from initial 400 μm slides by grinding sections using an grinding machine (KULZER EXAKT 400CS, EXAKT, Norderstedt, Germany). Hematoxylin-eosin staining was perfomed, and images were captured by computer connected to light microscope (Olympus BX, Olympus, Tokyo, Japan) attached to a CCD camera (Polaroid DMC2 digital Microscope camera (Polaroid Corporation, Cambridge, MA, USA). All assessments were made by one skilled investigator using SPOT Software (Ver. 4.0, Diagnostic Instrument, Inc., Sterling Heights, MI, USA).

The following parameters [36] were evaluated:

- Bone growth height in buccal defect areas (BG, mm): The thickness of bone that grew upward from the implant from the reference point on the buccal defect site on the alveolar ridge.
- Bone to implant contact in microthreads (microBIC, %): The bone to implant contact ratio was measured in buccal and lingual defect areas where the bone grew along the implant from the implantation reference point on the alveolar ridge.
- Bone to implant contact in macrothreads (macroBIC, %): The bone to implant contact ratio was measured in existing bone where the implant was implanted.
- Intra-thread bone density in macrothreads (ITBD, %): Intra-thread bone density was measured in the existing bone where the implant was placed.

After measuring the percentage of bone to implant contact (BIC, %), the ratio of bone formation area on intra-threads of implant to overall threads was calculated to determine intra-thread bone density (ITBD, %). Height of newly formed marginal bone by implants was measured. images of specimens were captured at a magnification of ×12.5 and ×40. For the histometric analysis, a magnification of ×40 was used.

2.5.7. Statistical Analysis

All of the analyses were performed using statistical program (SPSS ver. 21.0). Mean and SDs of ISQ values and of BIC and ITBD values were calculated for each group. Comparisons of ISQ values between the experimental and control groups were made using the Mann-Whitney U test. The Shapiro-Wilk test was used to test the normalities of distributions, and then one-way ANOVA was used to compare group BICs, ITBDs, and bone growths. *Post hoc* testing was performed using Bonferroni's test using a significance level of 95%.

3. Results

3.1. Characterization of Ti and Modified Ti Morpholgies

3.1.1. Scanning Electron Microscopy (SEM)

As shown in Figure 3, the surface morphologies of Ti, Hepa/Ti, PDGF/Hepa/Ti, BMP/Hepa/Ti, and PDGF/BMP/Hepa/Ti discs were investigated using SEM. Ti modified with rhHepa-DOPA, rhPDGF-BB, rhBMP-2, or rhPDGF-BB/rhBMP-2 had surface morphologies similar to Ti alone. These results indicate that the surfaces of Ti modified by small molecules, such as, Hepa-DOPA, rhPDGF-BB, rhBMP-2, or rhPDGF-BB/rhBMP-2 cannot be differentiated by SEM.

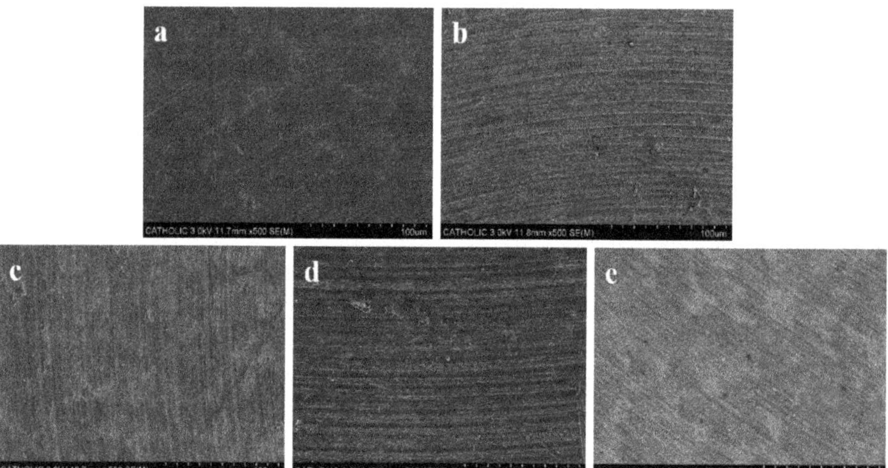

Figure 3. Scanning electron microscope (SEM) images of (**a**) Ti, (**b**) Hepa/Ti, (**c**) PDGF/Hepa/Ti, (**d**) BMP/Hepa/Ti, and (**e**) PDGF/BMP/Hepa/Ti.

3.1.2. X-ray Photoelectron Spectroscopy (XPS)

The surface chemical compositions of all Ti substrates determined by XPS are shown in Table 1. After anchoring Hepa-DOPA on the surface of Ti, C1s, and N1s peaks increased to compare with Ti alone. After immobilizing rhPDGF-BB and/or rhBMP-2 on the surface of Hepa/Ti disc, the N1s peak was increased and the S2p peak decreased versus Hepa/Ti. The amount of heparin anchored onto the Ti surface was 1.62 ± 0.32 μg/disc.

Table 1. Surface chemical compositions evaluated in 1 disc per group.

Specimen	C1s (%)	N1s (%)	O1s (%)	S2p (%)	Ti2p (%)	Total (%)
Ti	56.47	0.88	30.46	-	12.19	100
Hepa/Ti	62.01	3.01	28.63	0.56	5.79	100
PDGF/Hepa/Ti	60.86	5.28	30.23	0.36	3.27	100
BMP/Hepa/Ti	61.52	5.90	28.99	0.41	3.18	100
PDGF/BMP/Hepa/Ti	60.25	5.87	30.56	0.30	3.02	100

3.2. *In Vitro rhPDGF-BB and rhBMP-2 Releases*

The release behaviors of rhPDGF-BB or rhBMP-2 from PDGF/Hepa/Ti, BMP/Hepa/Ti, and PDGF/BMP/Hepa/Ti are shown in Figure 4, respectively. The amounts of rhPDGF-BB released from PDGF/Hepa/Ti and PDGF/BMP/Hepa/Ti discs were 25.20 ± 6.48 ng and 15.56 ± 4.55 ng after 1 day, respectively. Over 28 days, the amounts of rhPDGF-BB released were 66.69 ± 5.81 ng for PDGF/Hepa/Ti and 34.52 ± 4.55 ng for PDGF/BMP/Hepa/Ti. In addition, on day 1, the amounts of rhBMP-2 that is released from BMP/Hepa/Ti and PDGF/BMP/Hepa/Ti were 27.46 ± 6.71 ng and 16.56 ± 4.48 ng, respectively. Over the 28-day period, the amount of rhBMP-2 released was 69.85 ± 7.43 ng for BMP/Hepa/Ti and 37.52 ± 4.26 ng for PDGF/BMP/Hepa/Ti.

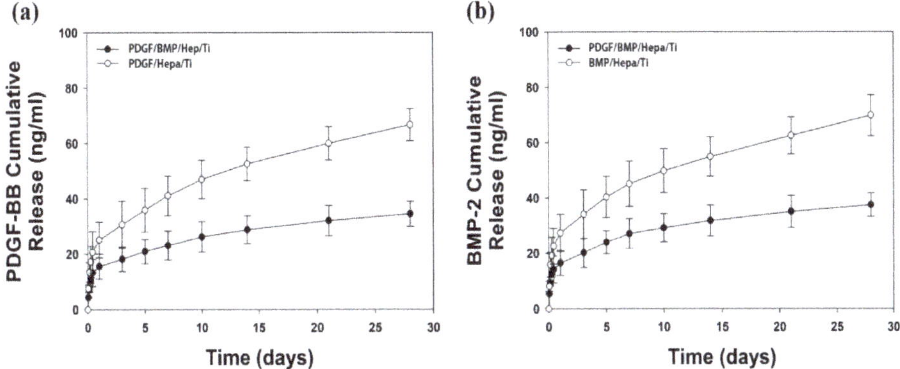

Figure 4. In vitro releases of (**a**) rhPDGF-BB and (**b**) rhBMP-2 from PDGF/Hepa/Ti, BMP/Hepa/Ti and PDGF/BMP/Hepa/Ti.

3.3. *In Vitro Cell Study*

3.3.1. ALP Activity

ALP activities of MG-63 cells seeded on the surface of Ti, PDGF/Hepa/Ti, BMP/Hepa/Ti, and PDGF/BMP/Hepa/Ti discs were confirmed after 3, 7, and 10 days of culture (Figure 5a). On day 3, the ALP activities of cells cultured on PDGF/Hepa/Ti, BMP/Hepa/Ti, and PDGF/BMP/Hepa/Ti were higher than that of cells cultured on Ti. On days 7 and 10, the ALP activities of cells that are grown on PDGF/Hepa/Ti, BMP/Hepa/Ti, and PDGF/BMP/Hepa/Ti differed significantly from those grown on Ti alone. On days 7 and 10, the ALP activity of cells cultivated on BMP/Hepa/Ti was significantly higher than that of those cultyivated on PDGF/Hepa/Ti or PDGF/BMP/Hepa/Ti.

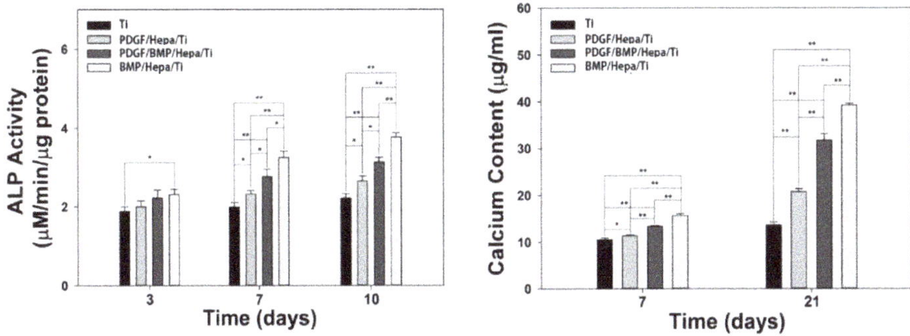

Figure 5. (**a**) Alkaline phosphatase (ALP) activities of cells cultured on Ti, PDGF/Hepa/Ti, BMP/Hepa/Ti, and PDGF/BMP/Hepa/Ti for 3, 7, or 10 days (* $p < 0.05$ and ** $p < 0.01$); (**b**) Calcium contents of cells grown on Ti, PDGF/Hepa/Ti, BMP/Hepa/Ti, and PDGF/BMP/Hepa/Ti for 7 or 21 days (* $p < 0.05$ and ** $p < 0.01$).

3.3.2. Calcium Contents

The calcium contents of MG-63 cells cultured on Ti, PDGF/Hepa/Ti, BMP/Hepa/Ti, and PDGF/BMP/Hepa/Ti discs for 7 and 21 days are shown in Figure 4. The calcium contents of cells cultivated on PDGF/Hepa/Ti, BMP/Hepa/Ti, and PDGF/BMP/Hepa/Ti were significantly greater than that those that were grown on Ti alone on days 7 and 21. Moreover, the calcium contents of cells grown on BMP/Hepa/Ti and PDGF/Hepa/Ti and on BMP/Hepa/Ti and PDGF/BMP/Hepa/Ti differed significantly.

3.3.3. Gene Expressions

After 7 and 21 days of culture, the mRNA expression levels of OCN and OPN in cells grown on Ti disc or on Ti modified with rhPDGF-BB and/or rhBMP-2 were determined by real-time PCR (Figure 6). mRNA expression levels of OCN and OPN in cells cultured on BMP/Hepa/Ti or PDGF/BMP/Hepa/Ti were markedly higher than in those cultured on Ti alone at day 7. In addition, OCN and OPN expressions in cells cultivated on PDGF/Hepa/Ti were greater than in those cultivated on Ti alone at day 7, but not significantly so. On day 21, the expression levels of OCN and OPN in cells that are grown on Hepa/Ti immobilized with rhPDGF-BB and/or rhBMP-2 and Ti alone were significantly different, as were OCN and OPN expressions in cells cultivated on BMP/Hepa/Ti and PDGF/Hepa/Ti or PDGF/BMP/Hepa/Ti discs.

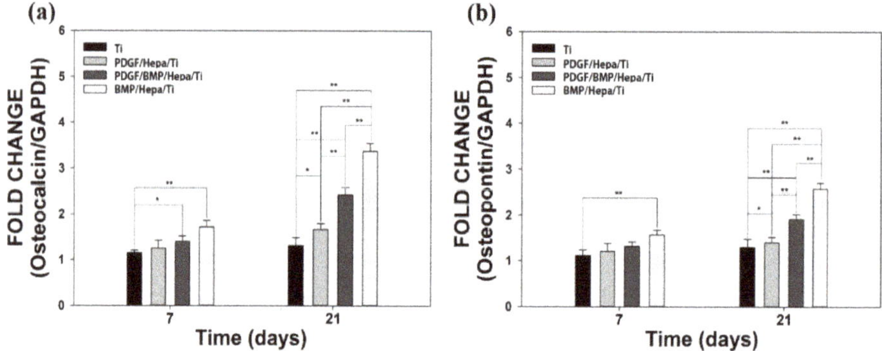

Figure 6. Real-time PCR analysis of the mRNA levels of (**a**) osteocalcin and (**b**) osteopontin in cells grown on Ti, PDGF/Hepa/Ti, BMP/Hepa/Ti, or PDGF/BMP/Hepa/Ti discs for 7 or 21 days (* $p < 0.05$ and ** $p < 0.01$).

3.4. *In Vivo Animal Study*

3.4.1. Clinical Findings

All of the experimental animals survived during the healing procedure, and there was no observation of infection or inflammation at surgical sites. Samples of 40 implants were collected without any issue for in vivo μCT and Histomorphometric analysis.

3.4.2. Stability Evaluation

Immediately after implantation, the ISQ values in all of the experimental groups were higher than in control group, but differences were not statistically significant. At eight weeks after surgery, ISQ values were higher than at baseline in the experimental groups, whereas in the control group, they were lower or similar than baseline values. ISQ increases were significantly greater in the experimental groups than in the control group ($p < 0.05$) (Table 2). But, no difference was observed between the groups ($p > 0.05$).

Table 2. Implant stability quotient (ISQ) values of groups immediately and 8 weeks after implantation.

Group	At Surgery	At 8 Weeks	ISQ Change
Ti	70.00 ± 4.45 [a]	71.27 ± 7.67 [a]	0.27 ± 7.97 [a]
Hepa/Ti	70.59 ± 8.17 [a]	70.70 ± 5.78 [a]	0.01 ± 9.19 [a]
PDGF/Hepa/Ti	69.99 ± 5.12 [a]	71.01 ± 4.98 [a]	0.92 ± 8.32 [a]
BMP/Hepa/Ti	71.33 ± 6.82 [a]	77.14 ± 5.23 [b]	5.32 ± 5.33 [b]
PDGF/BMP/Hepa/Ti	70.43 ± 6.44 [a]	81.41 ± 5.11 [b]	8.11 ± 7.09 [b]
* p	0.1255	0.0201	0.0011

Notes: At surgery: ISQ value at surgery; At week 8: ISQ value 8 weeks after surgery; [a,b]: Numbers in the same column with the same superscripts were not significantly different; p values were obtained by ANOVA; * $p < 0.05$.

3.4.3. Micro Computed Tomography (μCT)

Mean bone volume (%) in the control group was 18.19 ± 4.39% and in the Hepa/Ti, PDGF/Hepa/Ti, BMP/Hepa/Ti, and PDGF/BMP/Hepa/Ti groups mean bone volumes were 13.05 ± 5.15, 21.29 ± 9.56, 40.95 ± 9.07, and 43.73 ± 6.75%, respectively (Figure 7; n = 8). Bone volumes in the BMP/Hepa/Ti and PDGF/BMP/Hepa/Ti groups were significantly higher than in the other groups ($p < 0.05$). But, there was no significant difference between the BMP/Hepa/Ti and PDGF/BMP/Hepa/Ti groups ($p > 0.05$).

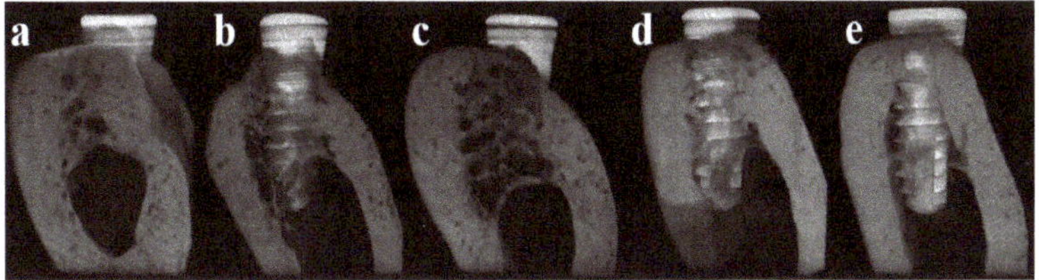

Figure 7. μCT 3D images of the five study groups. (**a**) Ti group; (**b**) Hepa/Ti group; (**c**) PDGF/Hepa/Ti group; (**d**) BMP/Hepa/Ti group; and (**e**) PDGF/BMP/Hepa/Ti group (×12.5 magnification). Note pronounced re-modeling of peri-implant bone and vertical bone growth in the BMP/Hepa/Ti and PDGF/Hepa/Ti groups. No bone growth was observed in defect areas in the Ti, Hepa/Ti, and PDGF/Hepa/Ti groups.

3.4.4. Histomorphometric Analysis

Jaw quadrants in the BMP/Hepa/Ti and PDGF/BMP/Hepa/Ti groups exhibited bone formation approaching microthreads and bone remodeling around implants (Figure 8). There was about 1.1 mm more bone gain in the BMP/Hepa/Ti and PDGF/BMP/Hepa/Ti groups than in the control group ($p < 0.05$). In the control and heparin groups, vertical bone loss was observed in buccal defect areas. The control, heparin, and PDGF groups showed failure of osseointegration in lingual regions observed a 1 mm gap between fixture and bone (Figure 9). The microBIC values of the BMP/Hepa/Ti and PDGF/BMP/Hepa/Ti groups were significantly greater than those of the other groups ($p < 0.05$), but no significant difference was observed between the BMP/Hepa/Ti and PDGF/BMP/Hepa/Ti groups ($p > 0.05$). The BMP/Hepa/Ti and PDGF/BMP/Hepa/Ti groups showed successful osseointegration in lingual portions (Figure 10). Mean group percentages (±SD) of BIC and ITBD within macrothreads eight weeks after surgery were not significantly different ($p > 0.05$) (Table 3).

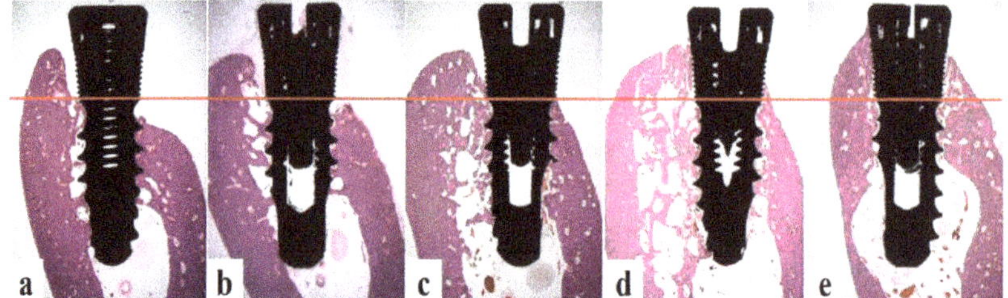

Figure 8. Histological specimens of the five groups. (**a**) Ti group; (**b**) Hepa/Ti group; (**c**) PDGF/Hepa/Ti group; (**d**) BMP/Hepa/Ti group; and (**e**) PDGF/BMP/Hepa/Ti group (×12.5 magnification). Note pronounced peri-implant bone re-modeling and vertical bone growth in the BMP/Hepa/Ti and PDGF/BMP/Hepa/Ti groups. Red lines indicate defect levels.

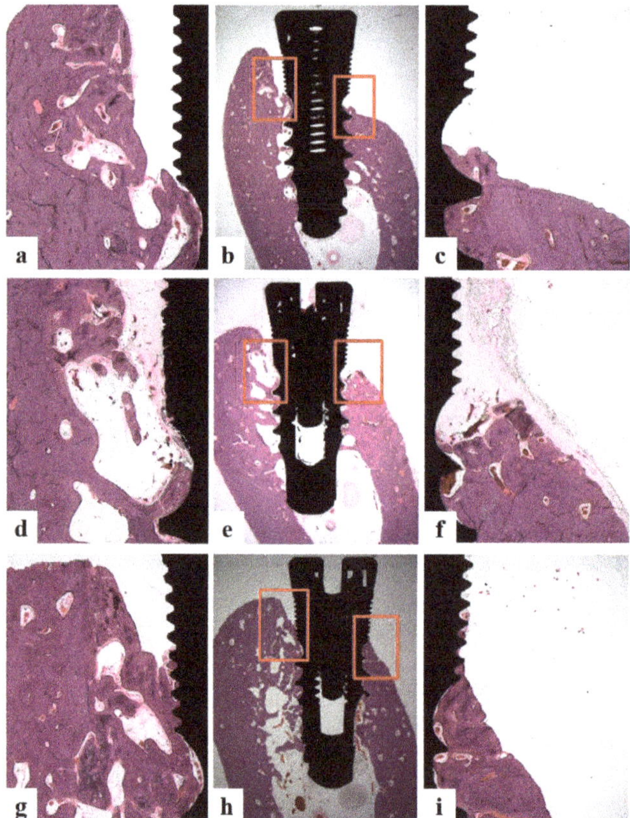

Figure 9. Microthread defect areas in the control (Ti), Hepa/Ti and PDGF/Hepa/Ti groups. Group BIC and ITBD within macrothreads at 8 weeks after surgery were not significantly different. In the control and heparin groups, vertical bone loss was detected in buccal defect areas. Failure of osseointegration in the lingual portion observed as a 1 mm gap between fixture and bone was present in the control, heparin, and PDGF groups. (**a–c**) Ti group; (**d–f**) Hepa/Ti group; (**g–i**) PDGF/Hepa/Ti group; (**a**,**d**,**g**) lingual portion of microthread; and (**c**,**f**,**i**) buccal portion of microthread. (**a**,**c**,**d**,**f**,**g**,**i**) ×40 magnification, (**b**,**e**,**h**) ×12.5 magnification.

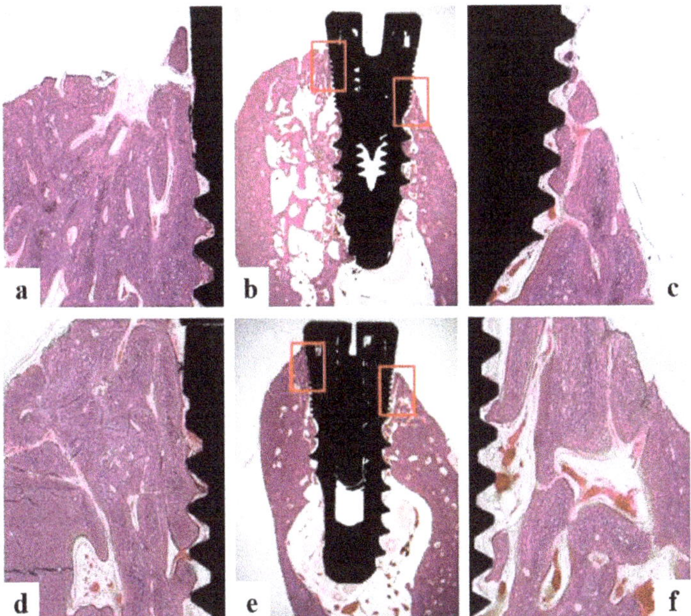

Figure 10. Microthread defect area in the BMP/Hepa/Ti and PDGF/BMP/Hepa/Ti groups. The BMP/Hepa/Ti and PDGF/BMP/Hepa/Ti group exhibited successful osseointegration in lingual portions. (**a**,**d**) lingual microthread portion; (**c**,**f**) buccal microthread portion; (**a**–**c**) BMP/Hepa/Ti group; (**d**–**f**) PDGF/BMP/Hepa/Ti group. (**a**,**c**,**d**,**f**) ×40 magnification, (**b**,**e**) ×12.5 magnification.

Table 3. Results of histometric analysis performed at 8 weeks after surgery. (Buccal bone level, BIC on macrothreads, and bone density on macrothreads; $n = 8$).

Group	BG (mm)	microBIC (%)	macroBIC (%)	ITBD (%)
Ti	0.23 ± 0.22^{a}	11.05 ± 5.09^{a}	23.58 ± 1.63^{a}	54.90 ± 7.24^{a}
Hepa/Ti	-0.06 ± 0.21^{a}	9.27 ± 1.95^{a}	18.47 ± 2.89^{a}	53.98 ± 3.77^{a}
PDGF/Hepa/Ti	0.12 ± 0.28^{a}	9.59 ± 3.99^{a}	20.62 ± 2.30^{a}	61.64 ± 6.17^{a}
BMP/Hepa/Ti	1.34 ± 0.17^{b}	27.76 ± 3.03^{b}	22.20 ± 2.89^{a}	60.80 ± 3.32^{a}
PDGF/BMP/Hepa/Ti	1.31 ± 0.12^{b}	31.79 ± 3.90^{b}	23.54 ± 2.30^{a}	69.22 ± 3.96^{a}
$^{*}p$	0.000	0.000	0.544	0.244

Notes: BG: Bone growth height in buccal defect area (mm); microBIC: Bone to implant contact in the microthreads (%); macroBIC: Bone to implant contact in macrothreads (existing bone area around implant, %); ITBD: Intra-thread bone density in macrothreads (existing bone area around implant, %); ^{a,b}: Means with the same superscripts in same columns were not significantly different; p values were obtained by ANOVA; $^{*}p < 0.05$.

4. Discussion

Many studies have been conducted on implant surfaces treated with growth factors like BMP-2, to increase vertical and horizontal bone regeneration without the use of additional bone grafts or barrier membranes [37,38]. Although BMP-2 is one of the effective growth factors, rapid release at an early stage and rapid diffusion into body fluids has limited its clinical applications [39,40]. In order to overcome these problems, the present study was undertaken on a heparin-based release system to provide the sustained and local release of BMP-2 [30,41,42]. Since titanium has good mechanical properties, corrosion resistance, and biocompatibility without biological activity, it is generally used as an implant material [43]. Due to its inertness in vivo, physical adsorption or chemical coatings are required to enable biomolecules to adhere to titanium surfaces [44]. However, the chemical substances used for this purpose, such as, APTES, EDAC, and NHS, are harmful to the human body [45]. In this study, titanium was coated with heparin using a dopamine surface modification technique, which has been reported to be biocompatible in in vivo toxicity studies [46]. Dopamine is a small molecule substance that possesses catechol and amine functional groups [47]. The catechol component bonds to titanium oxide surfaces and provides an amine group that can bind heparin [48,49].

It has been reported that combinations of different growth factors produce synergistic effects that promote the complex cellular activities that occur during bone regeneration and osseointegration [33,50,51]. However, most studies are limited to cell experiments, the studies reproducing the clinical situation in large animals are rare. Therefore, in the present study, rhBMP-2 and rhPDGF-BB immobilized on Ti surfaces modified with Hepa-DOPA were utilized to establish an effective growth factor delivery system to ensure sustained factor release in appropriate amounts over sufficient time in beagle mandible model. The concentration of rhBMP-2 that is used in this study was set based on our previous studies, which reported 0.75 mg/mL of rhBMP-2 is effective on bone regeneration [34,35]. In PDGF, the previous study of Choo et al. [33] reported that the appropriate concentration of PDGF was from 0.3 mg/mL up to 1 mg/mL, thus we chose 0.75 mg/mL of PDGF concentration within that range. BMP and PDGF combined groups were studied in 1:10 ratio (PDGF:BMP) in previous cell study [6], but we intended to increase the rate of PDGF to get enhanced bone regeneration as ratio of 1:1. The results of our in vitro release experiments indicated that the inclusion of heparin achieved sustained growth factor release, although previous studies have reported a burst release pattern resulting in the release of 70%–90% over the first 6 h [52]. After 24 h, the amounts of PDFG-BB and rhBMP released from PDGF/Hepa/Ti and BMP/Hepa/Ti were approximately 38%, respectively, and the amounts of rhBMP and PDGF-BB released from PDGF/BMP/Hepa/Ti were 45%, respectively. This suggests that PDGF-BB and rhBMP-2 show similar release tendencies in the presence of other growth factors and heparin.

SEM analysis of Ti surface modified by heparin, rhPDGF-BB, or/and rhBMP-2 showed their surface morphologies were similar to that of untreated Ti. XPS showed heparin was successfully grafted on anodized Ti surfaces, as indicated by higher C and N peaks. In a previous study, successful immobilization of rhBMP-2 on surface of carboxymethyl chitosan (CMCS)-grafted Ti was attributed to the covalent bonds formation between the carboxyl groups of CMCS and the amine groups of rhBMP-2 [53]. In the present study, successful anchoring of rhBMP-2 and rhPDGF-BB to Hepa/Ti surfaces was demonstrated by further increases in N content.

ALP activity and calcium deposition as determined by in vitro studies are widely used as markers of early and late differentiation to osteoblasts, respectively [54,55]. We measured ALP activity after culturing MG-63 for 3, 7, or 10 days and calcium deposition after 7 or 21 days. ALP activities of cells that are cultured on PDGF-BB and/or BMP/Hepa/Ti were found to be significantly higher than those of cells cultured on untreated Ti on days 7 and 10 (* $p < 0.05$ and ** $p < 0.01$). As reported in a previous study [6,56], our finding confirmed that rhBMP-2 and rhPDGF-BB both stimulate osteoblast differentiation. Furthermore, the calcium contents of cells cultured on PDGF-BB or/and BMP/Hepa/Ti were significantly higher than those of cells cultured on untreated Ti on days 7 and 21 (* $p < 0.05$ and ** $p < 0.01$). These results indicate that growth factor immobilized Ti substrates can stimulate matrix formation and improve osteoblast cell function. ALP activity and calcium deposition in the PDGF/BMP/Hepa/Ti group were significantly higher than in the PDGF/Hepa/Ti group, but were significantly lower than in the BMP/Hepa/Ti group. In order to investigate the gene expression levels of osteocalcin (OCN) and osteopontin (OPN), which are markers of cell differentiation, the total mRNAs of cells cultured on four Ti surfaces were extracted. On day 21, significant differences were observed between the OCN and OPN expressions of cells cultivated on BMP/Hepa/Ti and PDGF/Hepa/Ti or PDGF/BMP/Hepa/Ti.

In the result of previous cell studies [6], PDGF-BB (5 ng/mL)/BMP-2(50 ng/mL)/Hepa/Ti (a 1:10 ratio of growth factors) was significantly higher than BMP/Hepa/Ti, whereas in this study, PDGF-BB

(50 ng/mL)/BMP-2 (50 ng/mL)/Hepa/Ti (a 1:1 ratio) was significantly lower than BMP/Hepa/Ti. Furthermore, ALP, calcium deposition, and gene expression not synergistically increased by the copresence of rhPDGF-BB and rhBMP-2. These results indicate that different concentrations and ratios of PDGF-BB should be selected.

The in vivo animal study was conducted to evaluate the synergic effects of rhBMP-2 (0.75 mg/mL) when combined with rhPDGF-BB (0.75 mg/mL). A total of 40 implants of five groups (control (Ti) group, Hepa/Ti group, PDGF/Hepa/Ti group, BMP/Hepa/Ti group, and PDGF/BMP/Hepa/Ti group) were implanted without using additional implants or barrier membranes. To reproduce clinical situations, a buccal open defect model with a 2.5 mm deep peri-implant bone defect was produced in the mandibles of five beagle dogs. This model was designed with loss of buccal bone and a mesial-lingual-distal 1 mm defect area around the 2.5 mm upper portions of implants. The upper part of implants (microthread) was exposed 2.5 mm above alveolar bone and the lower part (macrothread) was implanted into cortical bone. After a healing period of eight weeks, ISQ values showed the presences of rhBMP-2 and rhPDGF-BB contributed positively to implant stability. Furthermore, ISQ increases were significantly higher in the PDGF/BMP/Hepa/Ti and BMP/Hepa/Ti groups than in the other three groups (* $p < 0.05$). However, there was no difference between these two experimental groups ($p > 0.05$). μCT analysis showed new bone volumes in the BMP/Hepa/Ti and PDGF/BMP/Hepa/Ti groups were significantly greater than in the other groups (* $p < 0.05$), but no significant difference was observed between these two groups ($p > 0.05$). Our histomorphometric analysis confirmed vertical new bone formation and osseointegration approaching microthreads around peri-implant bone defects in the BMP/Hepa/Ti and PDGF/BMP/Hepa/Ti groups, whereas osseointegration was unsuccessful in the Ti, Hepa/Ti and PDGF/Hepa/Ti groups, in which a 1 mm gap between implant fixture and bone was observed at lingual portions. microBIC values in the BMP/Hepa/Ti and PDGF/BMP/Hepa/Ti groups were significantly greater than in the other groups (* $p = 0.000$), but no significant difference was observed between the two ($p > 0.05$). Our in vivo studies were confirmed that PDGF (0.75 mg/mL)/BMP (0.75 mg/mL)/Hepa/Ti (a 1:1 growth factor) had no synergic effect. Based on these results and limitations of present study, further studies on subdivided concentrations and ratios, various implant surfaces and slow releasing systems are required to more comprehensively investigate the effects of combinations of different growth factors on osseointegration and bone regeneration.

5. Conclusions

Anodized Ti surfaces were successfully functionalized by surface grafting heparin and subsequently immobilizing rhBMP-2 and/or rhPDGF-BB. rhBMP-2 immobilized, heparin-grafted Ti implants were found to provide a suitable delivery system that enhanced osseointegration and bone regeneration in defect areas around implants. However, the study failed to reveal any synergetic effect resulting from the co-immobilization of rhBMP-2 and rhPDGF-BB.

Acknowledgments: This study was supported by a National Research Foundation of Korea (NRF) grant funded by the Korea government (MSIP) (No. 2017R1A2B4005820).

Author Contributions: Jung-Bo Huh conceived and designed the experiments; So-Hyoun Lee, Eun-Bin Bae, Sung-Eun Kim, Young-Pil Yun, Hak-Jun Kim, Jae-Won Choi and Jin-Ju Lee performed the experiments; So-Hyoun Lee, Sung-Eun Kim, Young-Pil Yun, Hak-Jun Kim and Jung-Bo Huh analyzed the data; So-Hyoun Lee, Eun-Bin Bae, and Jung-Bo Huh wrote the paper.

Conflicts of Interest: The authors have no conflict of interest to declare.

References

1. Poli, P.P.; Beretta, M.; Cicciù, M.; Maiorana, C. Alveolar ridge augmentation with titanium mesh. A retrospective clinical study. *Open Dent. J.* **2014**, *8*, 148–158. [PubMed]
2. Retzepi, M.; Lewis, M.P.; Donos, N. Effect of diabetes and metabolic control on de novo bone formation following guided bone regeneration. *Clin. Oral Implants Res.* **2010**, *21*, 71–79. [CrossRef] [PubMed]

3. Rabel, A.; Köhler, S.G.; Schmidt-Westhausen, A.M. Clinical study on the primary stability of two dental implant systems with resonance frequency analysis. *Clin. Oral Investig.* **2007**, *11*, 257–265. [CrossRef] [PubMed]
4. Chen, D.; Zhao, M.; Mundy, G.R. Bone morphogenetic proteins. *Growth Factors* **2004**, *22*, 233–241. [CrossRef] [PubMed]
5. Hall, J.; Sorensen, R.G.; Wozney, J.M.; Wikesjo, U.M. Bone formation at rhBMP-2-coated titanium implants in the rat ectopic model. *J. Clin. Periodontol.* **2007**, *34*, 444–451. [CrossRef] [PubMed]
6. Kim, S.E.; Yun, Y.P.; Lee, J.Y.; Shim, J.S.; Park, K.; Huh, J.B. Co-delivery of platelet-derived growth factor (PDGF-BB) and bone morphogenic protein (BMP-2) coated onto heparinized titanium for improving osteoblast function and osteointegration. *J. Tissue Eng. Regen. Med.* **2015**, *9*, E219–E228. [CrossRef] [PubMed]
7. Bessa, P.C.; Casal, M.; Reis, R.L. Bone morphogenetic proteins in tissue engineering: The road from laboratory to clinic, part II (BMP delivery). *J. Tissue Eng. Regen. Med.* **2008**, *2*, 81–96. [CrossRef] [PubMed]
8. Bessa, P.C.; Casal, M.; Reis, R.L. Bone morphogenetic proteins in tissue engineering: The road from the laboratory to the clinic, part I (basic concepts). *J. Tissue Eng. Regen. Med.* **2008**, *2*, 1–13. [CrossRef] [PubMed]
9. Becker, J.; Kirsch, A.; Schwarz, F.; Chatzinikolaidou, M.; Rothamel, D.; Lekovic, V.; Jennissen, H.P. Bone apposition to titanium implants biocoated with recombinant human bone morphogenetic protein-2 (rhBMP-2). A pilot study in dogs. *Clin. Oral Investig.* **2006**, *10*, 217–224. [CrossRef] [PubMed]
10. Wikesjo, U.M.E.; Xiropaidis, A.V.; Qahash, M.; Lim, W.H.; Sorensen, R.G.; Rohrer, M.D.; Wozney, M.; Hall, J. Bone formation at recombinant human bone morphogenetic protein-2-coated titanium implants in the posterior mandible (Type II bone) in dogs. *J. Clin. Periodontol.* **2008**, *35*, 985–991. [CrossRef] [PubMed]
11. Stenport, V.F.; Johansson, C.; Heo, S.J.; Aspenberg, P.; Albrektsson, T. Titanium implants and BMP-7 in bone: An experimental model in the rabbit. *J. Mater. Sci. Mater. Med.* **2003**, *14*, 247–254. [CrossRef] [PubMed]
12. Schliephake, H.; Aref, A.; Scharnweber, D.; Bierbaum, S.; Roessler, S.; Sewing, A. Effect of immobilized bone morphogenic protein 2 coating of titanium implants on peri-implant bone formation. *Clin. Oral Implants Res.* **2005**, *16*, 563–569. [CrossRef] [PubMed]
13. Park, J.; Lutz, R.; Felszeghy, E.; Wiltfang, J.; Nkenke, E.; Neukam, F.W.; Schlegel, K.A. The effect on bone regeneration of a liposomal vector to deliver BMP-2 gene to bone grafts in peri-implant bone defects. *Biomaterials* **2007**, *28*, 2772–2782. [CrossRef] [PubMed]
14. Stadlinger, B.; Pilling, E.; Huhle, M.; Mai, R.; Bierbaum, S.; Scharnweber, D.; Kuhlisch, E.; Loukota, R.; Eckelt, U. Evaluation of osseointegration of dental implants coated with collagen, chondroitin sulphate and BMP-4: An animal study. *Int. J. Oral Maxillofac. Surg.* **2008**, *37*, 54–59. [CrossRef] [PubMed]
15. Ortolani, E.; Guerriero, M.; Coli, A.; Di Giannuario, A.; Minniti, G.; Polimeni, A. Effect of PDGF, IGF-1 and PRP on the implant osseointegration. A histological and immunohistochemical study in rabbits. *Ann. Stomatol. (Roma)* **2014**, *5*, 66–68. [CrossRef] [PubMed]
16. Chung, S.M.; Jung, I.K.; Yoon, B.H.; Choi, B.R.; Kim, D.M.; Jang, J.S. Evaluation of Different Combinations of Biphasic Calcium Phosphate and Growth Factors for Bone Formation in Calvarial Defects in a Rabbit Model. *Int. J. Periodontics Restor. Dent.* **2016**, *36*, s49–s59. [CrossRef] [PubMed]
17. Deshpande, A.; Koudale, S.B.; Bhongade, M.L. A comparative evaluation of rhPDGF-BB + β-TCP and subepithelial connective tissue graft for the treatment of multiple gingival recession defects in humans. *Int. J. Periodontics Restor. Dent.* **2014**, *34*, 241–249. [CrossRef] [PubMed]
18. Ridgway, H.K.; Mellonig, J.T.; Cochran, D.L. Human histologic and clinical evaluation of recombinant human platelet-derived growth factor and beta-tricalcium phosphate for the treatment of periodontal intraosseous defects. *Int. J. Periodontics Restor. Dent.* **2008**, *28*, 171–179.
19. Thakare, K.; Deo, V. Randomized controlled clinical study of rhPDGF-BB + β-TCP versus HA + β-TCP for the treatment of infrabony periodontal defects: Clinical and radiographic results. *Int. J. Periodontics Restor. Dent.* **2012**, *32*, 689–696.
20. Mishra, A.; Mishra, H.; Pathakota, K.R.; Mishra, J. Efficacy of modified minimally invasive surgical technique in the treatment of human intrabony defects with or without use of rhPDGF-BB gel: A randomized controlled trial. *J. Clin. Periodontol.* **2013**, *40*, 172–179. [CrossRef] [PubMed]
21. Deschaseaux, F.; Sensebe, L.; Heymann, D. Mechanisms of bone repair and regeneration. *Trends Mol. Med.* **2009**, *15*, 417–429. [CrossRef] [PubMed]

22. Misch, C.M.; Nevins, M.L.; Boch, J.A. Recombinant Human Bone Morphogenetic Protein-2 or Recombinant Human Platelet-Derived Growth Factor BB in Extraction Site Preservation and Bone Augmentation. In *Minimally Invasive Dental Implant Surgery*; Wiley-Blackwell: Hoboken, NJ, USA, 2015; pp. 119–136.
23. Pountos, I.; Georgouli, T.; Henshaw, K.; Bird, H.; Giannoudis, P.V. Release of growth factors and the effect of age, sex, and severity of injury after long bone fracture. *Acta Orthop.* **2013**, *84*, 65–70. [CrossRef] [PubMed]
24. Caplan, A.I.; Correa, D. PDGF in bone formation and regeneration: New insights into a novel mechanism involving MSCs. *J. Orthop. Res.* **2011**, *29*, 1795–1803. [CrossRef] [PubMed]
25. Centrella, M.; McCarthy, T.L.; Kusmik, W.F.; Canalis, E. Relative binding and biochemical effects of heterodimeric and homodimeric isoforms of platelet-derived growth factor in osteoblast-enriched cultures from fetal rat bone. *J. Cell. Physiol.* **1991**, *147*, 420–426. [CrossRef] [PubMed]
26. Hollinger, J.O.; Onikepe, A.O.; MacKrell, J.; Einhorn, T.; Bradica, G.; Lynch, S.; Hart, C.E. Accelerated fracture healing in the geriatric, osteoporotic rat with recombinant human platelet-derived growth factor-bb and an injectable beta-tricalcium phosphate/collagen matrix. *J. Orthop. Res.* **2008**, *26*, 83–90. [CrossRef] [PubMed]
27. Mitlak, B.H.; Finkelman, R.D.; Hill, E.L.; Li, J.; Martin, B.; Smith, T.; D'Andrea, M.; Antoniades, H.N.; Lynch, S.E. The effect of systemically administered PDGF-BB on the rodent skeleton. *J. Bone Miner. Res.* **1996**, *11*, 238–247. [CrossRef] [PubMed]
28. Martino, M.M.; Briquez, P.S.; Ranga, A.; Lutolf, M.P.; Hubbell, J.A. Heparin-binding domain of fibrin (ogen) binds growth factors and promotes tissue repair when incorporated within a synthetic matrix. *Proc. Natl. Acad. Sci. USA* **2013**, *110*, 4563–4568. [CrossRef] [PubMed]
29. Perets, A.; Baruch, Y.; Weisbuch, F.; Shoshany, G.; Neufeld, G.; Cohen, S. Enhancing the vascularization of three-dimensional porous alginate scaffolds by incorporating controlled release basic fibroblast growth factor microspheres. *J. Biomed. Mater. Res. A* **2003**, *65*, 489–497. [CrossRef] [PubMed]
30. Ishibe, T.; Goto, T.; Kodama, T.; Miyazaki, T.; Kobayashi, S.; Takahashi, T. Bone formation on apatite-coated titanium with incorporated BMP-2/heparin in vivo. *Oral Surg. Oral Med. Oral Pathol. Oral Radiol. Endod.* **2009**, *108*, 867–8975. [CrossRef] [PubMed]
31. Lee, S.H.; Jo, J.Y.; Yun, M.J.; Jeon, Y.C.; Huh, J.B.; Jeong, C.M. Effect of immobilization of the recombinant human bone morphogenetic protein 2 (rhBMP-2) on anodized implants coated with heparin for improving alveolar ridge augmentation in beagle dogs: Radiographic observations. *J. Korean Acad. Prosthodont.* **2013**, *51*, 307–314. [CrossRef]
32. Huh, J.B.; Lee, J.Y.; Lee, K.L.; Kim, S.E.; Yun, M.J.; Sim, J.S.; Shim, J.S.; Shin, S.W. Effects of the immobilization of heparin and rhPDGF-BB to titanium surfaces for the enhancement of osteoblastic functions and anti-inflammation. *J. Adv. Prosthodont.* **2011**, *3*, 152–160. [CrossRef] [PubMed]
33. Choo, T.; Marino, V.; Bartold, P.M. Effect of PDGF-BB and beta-tricalcium phosphate (β-TCP) on bone formation around dental implants: A pilot study in sheep. *Clin. Oral Implants Res.* **2013**, *24*, 158–166. [CrossRef] [PubMed]
34. Huh, J.B.; Park, C.K.; Kim, S.E.; Shim, K.M.; Choi, K.H.; Kim, S.J.; Sim, J.S.; Shin, S.W. Alveolar ridge augmentation using anodized implants coated with Escherichia coli–derived recombinant human bone morphogenetic protein 2. *Oral Surg. Oral Med. Oral Pathol. Oral Radiol. Endod.* **2011**, *112*, 42–49. [CrossRef] [PubMed]
35. Huh, J.B.; Kim, S.E.; Kim, H.E.; Kang, S.S.; Choi, K.H.; Jeong, C.M.; Lee, J.Y.; Shin, S.W. Effects of anodized implants coated with Escherichia coli-derived rhBMP-2 in beagle dogs. *Int. J. Oral Maxillofac. Surg.* **2012**, *41*, 1577–1584. [CrossRef] [PubMed]
36. Kim, S.E.; Kim, C.S.; Yun, Y.P.; Yang, D.H.; Park, K.S.; Kim, S.E.; Jeong, C.M.; Huh, J.B. Improving osteoblast functions and bone formation upon BMP-2 immobilization on titanium modified with heparin. *Carbohydr. Polym.* **2014**, *114*, 123–132. [CrossRef] [PubMed]
37. Jung, R.E.; Windisch, S.I.; Eggenschwiler, A.M.; Thoma, D.S.; Weber, F.E.; Hämmerle, C.H. A randomized-controlled clinical trial evaluating clinical and radiological outcomes after 3 and 5 years of dental implants placed in bone regenerated by means of GBR techniques with or without the addition of BMP-2. *Clin. Oral Implants Res.* **2009**, *20*, 660–666. [CrossRef] [PubMed]
38. Jung, R.E.; Glauser, R.; Schärer, P.; Hämmerle, C.H.; Sailer, H.F.; Weber, F.E. Effect of rhBMP-2 on guided bone regeneration in humans. *Clin. Oral Implants Res.* **2003**, *14*, 556–568. [CrossRef] [PubMed]

39. Boyne, P.J.; Lilly, L.C.; Marx, R.E.; Moy, P.K.; Nevins, M.; Spagnoli, D.B.; Triplett, R.G. De novo bone induction by recombinant human bone morphogenetic protein-2 (rhBMP-2) in maxillary sinus floor augmentation. *J. Oral Maxillofac. Surg.* **2005**, *63*, 1693–1707. [CrossRef] [PubMed]
40. Zhao, B.; Katagiri, T.; Toyoda, H.; Takada, T.; Yanai, T.; Fukuda, T.; Chung, U.I.; Koike, T.; Takaoka, K.; Kamijo, R. Heparin potentiates the in vivo ectopic bone formation induced by bone morphogenetic protein-2. *J. Biol. Chem.* **2006**, *281*, 23246–23253. [CrossRef] [PubMed]
41. Ho, Y.C.; Mi, F.L.; Sung, H.W.; Kuo, P.L. Heparin-functionalized chitosanealginate scaffolds for controlled release of growth factor. *Int. J. Pharm.* **2009**, *376*, 69–75. [CrossRef] [PubMed]
42. Lin, H.; Zhao, Y.; Sun, W.; Chen, B.; Zhang, J.; Zhao, W.; Xiao, Z.; Dai, J. The effect of crosslinking heparin to demineralized bone matrix on mechanical strength and specific binding to human bone morphogenetic protein-2. *Biomaterials* **2008**, *29*, 1189–1197. [CrossRef] [PubMed]
43. Le Guéhennec, L.; Soueidan, A.; Layrolle, P.; Amouriq, Y. Surface treatments of titanium dental implants for rapid osseointegration. *Dent. Mater.* **2007**, *23*, 844–854. [CrossRef] [PubMed]
44. Zhoua, D.; Ito, Y. Inorganic material surfaces made bioactive by immobilizing growth factors for hard tissue engineering. *RSC Adv.* **2013**, *3*, 11095–11106. [CrossRef]
45. Kim, S.E.; Song, S.H.; Yun, Y.P.; Choi, B.J.; Kwon, I.K.; Bae, M.S.; Moon, H.J.; Kwon, Y.D. The effect of immobilization of heparin and bone morphogenic protein-2 (BMP-2) to titanium surfaces on inflammationand osteoblast function. *Biomaterials* **2011**, *32*, 366–373. [CrossRef] [PubMed]
46. Hong, S.; Kim, K.Y.; Wook, H.J.; Park, S.Y.; Lee, K.D.; Lee, D.Y.; Lee, H. Attenuation of the in vivo toxicity of biomaterials by polydopamine surface modification. *Nanomedicine (London)* **2011**, *6*, 793–801. [CrossRef] [PubMed]
47. Lee, H.; Dellatore, S.M.; Miller, W.M.; Messersmith, P.B. Mussel-inspired surface chemistry for multifunctional coatings. *Science* **2007**, *318*, 426–430. [CrossRef] [PubMed]
48. Fan, X.; Lin, L.; Dalsin, J.L.; Messersmith, P.B. Biomimetic anchor for surface-initiated polymerization from metal substrates. *J. Am. Chem. Soc.* **2005**, *127*, 15843–15847. [CrossRef] [PubMed]
49. Lee, D.W.; Yun, Y.P.; Park, K.; Kim, S.E. Gentamicin and bone morphogenic protein-2 (BMP-2)-delivering heparinized-titanium implant with enhanced antibacterial activity and osteointegration. *Bone* **2012**, *50*, 974–982. [CrossRef] [PubMed]
50. Chaudhary, L.R.; Hofmeister, A.M.; Hruska, K.A. Differential growth factor control of bone formation through osteoprogenitor differentiation. *Bone* **2004**, *34*, 402–411. [CrossRef] [PubMed]
51. Zhang, Y.; Shi, B.; Li, C.; Wang, Y.; Chen, Y.; Zhang, W.; Luo, T.; Cheng, X. The synergetic bone-forming effects of combinations of growth factors expressed by adenovirus vectors on chitosan/collagen scaffolds. *J. Control. Release* **2009**, *136*, 172–178. [CrossRef] [PubMed]
52. Kim, N.H.; Lee, S.H.; Ryu, J.J.; Choi, K.H.; Huh, J.B. Effects of rhBMP-2 on sandblasted and acid etched titanium implant surfaces on bone regeneration and osseointegration: Spilt-mouth designed pilot study. *Biomed. Res. Int.* **2015**, *2015*, 1–11. [CrossRef] [PubMed]
53. Shi, Z.; Neoh, K.G.; Kang, E.T.; Poh, C.K.; Wang, W. Surface functionalization of titanium with carboxymethyl chitosan and immobilized bone morphogenetic protein-2 for enhanced osseointegration. *Biomacromolecules* **2009**, *10*, 1603–1611. [CrossRef] [PubMed]
54. Turksen, K.; Bhargava, U.; Moe, H.K.; Aubin, J.E. Isolation of monoclonal antibodies recognizing rat bone-associated molecules in vitro and in vivo. *J. Histochem. Cytochem.* **1992**, *40*, 1339–1352. [CrossRef] [PubMed]
55. Van den Beucken, J.J.; Walboomers, X.F.; Boerman, O.C.; Vos, M.R.; Sommerdijk, N.A.; Hayakawa, T.; Fukushima, T.; Okahata, Y.; Nolte, R.J.; Jansen, J.A. Functionalization of multilayered DNA-coatings with bone morphogenetic protein 2. *J. Control. Release* **2006**, *113*, 63–72. [CrossRef] [PubMed]
56. Skodje, A.; Idris, S.B.M.; Sun, Y.; Bartaula, S.; Mustafa, K.; Finne-Wistrand, A.; Wik, U.M.E.; Leknes, K.N. Biodegradable polymer scaffolds loaded with low-dose BMP-2 stimulate periodontal ligament cell differentiation. *J. Biomed. Mater. Res. A* **2015**, *103*, 1991–1998. [CrossRef] [PubMed]

Article

Application of Solution Plasma Surface Modification Technology to the Formation of Thin Hydroxyapatite Film on Titanium Implants

Akashlynn Badruddoza Dithi [1,2,*], Takashi Nezu [1], Futami Nagano-Takebe [1], Md Riasat Hasan [2], Takashi Saito [2] and Kazuhiko Endo [1]

1 Division of Biomaterials and Bioengineering, Department of Oral Rehabilitation, School of Dentistry, Health Sciences University of Hokkaido, 061-0293 Hokkaido, Japan; tnezu@hoku-iryo-u.ac.jp (T.N.); nagano23@hoku-iryo-u.ac.jp (F.N.-T.); endo@hoku-iryo-u.ac.jp (K.E.)

2 Division of Clinical Cariology and Endodontology, Department of Oral Rehabilitation, School of Dentistry, Health Sciences University of Hokkaido, 061-0293 Hokkaido, Japan; riasat@hoku-iryo-u.ac.jp (M.R.H.); t-saito@hoku-iryo-u.ac.jp (T.S.)

* Correspondence: akashlynn@hoku-iryo-u.ac.jp; Tel./Fax: +81-0133-23-1726

Received: 21 September 2018; Accepted: 13 December 2018; Published: 21 December 2018

Abstract: Hydroxyapatite (HA) coatings on titanium implants enhance rapid bone formation around the implant due to their osteoconductive property. The present study aimed to achieve a thin and uniform HA film coating on titanium implants by solution plasma treatment (SPT). Commercially pure titanium and porous titanium disks were employed. A pulse plasma generator was used on the disks for 30 min. Morphologic and crystallographic features of the deposited films were examined by scanning electron microscopy (SEM) and X-ray diffractometry (XRD). To evaluate the wettability of the disks, water droplet (20 μL) surfaces were measured using a contact angle analyzer. The initial attachment of osteoblast-like cells (MC3T3E1) on the titanium substrates before and after solution plasma treatment was evaluated by counting the number of attached cells after incubation for 4 h. After immersion in the mineralizing solution for up to seven days, no crystals were observed on the polished-Ti surface. A more uniform and dense precipitation of round and grown crystals with diameters of approximately 1–5 μm was observed on Ti-SPT. XRD clearly showed that the precipitated crystals on titanium disks were HA. The contact angle of the polished-Ti increased with time ($\theta = 37°–51°$). The surface of the Ti-SPT remained hydrophilic ($\theta < 5°$) after up to 30 days of aging. The number of attached cells on the Ti-SPT after aging for 30 days remained above 85% of that on the Ti-SPT without aging. SPT in a mineralizing solution can be used to acquire a homogenous precipitation of HA on porous-surfaced titanium implants.

Keywords: hydroxyapatite; titanium implants; mineralizing solution; solution plasma treatment

1. Introduction

The use of dental implants has revolutionized the current treatment of partially and fully edentulous patients given their high level of predictability and wide variety of treatment options. In 2006, the estimated number of dental titanium implants placed in the United States was over five million [1]. A major consideration in designing dental implants is the production of surfaces that promote desirable responses in the cells and tissues.

Hydroxyapatite (HA) coatings on titanium implants facilitate rapid bone formation due to their excellent osteoconductive property. HA-coated implants have been reported to stimulate bone healing, which enhances improvement in the rate and strength of the initial implant integration. For enhanced implant stability and bio-integration to bone tissue, various methods have been applied to coat the titanium implant with HA, for example, plasma spraying [2], electrophoretic co-deposition [3], ion-beam-sputter deposition [4], dip coating in a simulated body fluid [5], electrochemical deposition [6], blast coating [7], and thermally induced liquid-phase deposition [8,9]. Among these methods, titanium implant bodies coated with HA in dry processes, such as the plasma spraying and ion-beam-sputter deposition methods, have already been used clinically. The HA coating methods in dry processes, however, cannot be applied to titanium implant bodies with porous surface structures. To coat titanium implants with a three-dimensional porous surface structure with HA film, Kuroda et al. and Tamura et al. developed the titanium substrate heating method in a liquid [8,9]. Since the solubility product of HA decreases with an increase in temperature above 16 °C, they electrically heated the titanium substrate by applying a large current with an AC or DC power source to the substrate in calcium phosphate solutions to precipitate HA crystals on the titanium surface. These two studies demonstrated that the precipitated HA crystals were needle-like or plate-like in shape and that the formation of a uniform and dense HA film consisting of fine and spherical HA crystals could not be achieved. Another shortcoming of the titanium substrate heating method is that it requires a large electric power source exceeding 1000 V for the deposition of HA crystals on the surface of the dental implant body by Joule heating.

The solution plasma technique is an entirely new technology to coat an implant surface with HA spherical particles. To date, this solution plasma technique has been used in the fields of chemistry and applied physics [10]. When a pulsed voltage is applied between the two electrodes, a glowing discharge takes place in the liquid, which generates plasma in the gas phase. This plasma produces thermal energy and light, along with reactive chemical species such as hydrogen and hydroxide (OH) ions. With this novel surface modification technique, the temperature of the solution in contact with the titanium surface can easily be increased with the thermal energy generated by the solution plasma. In this case, a thin and uniform HA layer is expected to be formed on a porous-surfaced titanium implant, which will induce rapid bone growth into the pore space of the implant body, and also enhance the biological anchorage of the implant with bone.

The present study aimed to achieve a thin and uniform HA layer covering titanium with a porous surface structure by using solution plasma technology. The mechanism of HA precipitation on the titanium during solution plasma treatment was elucidated.

2. Materials and Methods

2.1. Substrate

Commercially pure titanium and porous titanium with 69.7% porosity were employed in this study.

Group A: Non-porous-Ti (Ti) (JIS Type-2, J. Morita Corp., Tokyo, Japan); spherical shape; specimen size was 14 mm in diameter and 2 mm in thickness.

Group B: Porous-Ti (powder sintering product with 69.7% porosity; Nagamine Manufacturing Company, Kagawa, Japan); specimen size was 14 mm^2, 1 mm in thickness, and the pore size was 287 ± 87 μm^2.

The Ti disks were polished with silicon carbide abrasive papers (#240, then #600). The polished groups were named "Polished-Ti". "Porous-Ti" was used without polishing for the experiments.

Half of the polished-Ti and porous-Ti substrates were subjected to alkaline treatment prior to the solution plasma treatment (SPT). The disks were immersed in 5 M NaOH solution at 60 °C for 24 h [11]. The alkaline treatment formed a micrometric network layer of sodium titanate hydrogel on the titanium surface, which enhanced the apetite nucleation in mineralizing solution.

After alkaline treatment, the polished titanium (Ti) group was named "Ti-AT" and the porous titanium substrate (porous-Ti) group was named "Porous-Ti-AT".

2.2. Preparation of the Mineralizing Solution

A metastable calcium phosphate solution was used in this study [12]. The composition of the solution is shown in Table 1. The Ca/P molar ratio of the mineralizing solution was 1.67, and it also contained 10 mM HEPES (*N*-2-Hydroxyethylpiperazine-*N*′-2-Ethanesulfonic acid) KOH (potassium hydroxide) for buffering at pH 7.4. The final ionic strength of this solution was adjusted to 0.16 by adding 150 mM KCl (potassium chloride). The preparation of the solution was completed by dissolving reagent grade $CaCl_2$ (calcium chloride), KH_2PO_4 (potassium phosphate), KCl (potassium chloride), and HEPES (*N*-2-Hydroxyethylpiperazine-*N*′-2-Ethanesulfonic acid) into deionized distilled water. The degree of supersaturation with respect to hydroxyapatite was 3.85×10^7 at 37 °C.

Table 1. Composition of the mineralizing solution.

Component	Concentration (mM)
$CaCl_2$	2.24
K_2HPO_4	1.04
KH_2PO_4	0.30
HEPES	10.00
KCl	150.00
C/P	1.67
pH	7.40
Degree of supersaturation with respect to HA	3.85×10^7

2.3. Solution Plasma Treatment

A solution plasma treatment was performed in 150 mL of the mineralizing solution at 29 °C. A pulse plasma generator (MPP-NV04, Pekuris, Tokyo, Japan) was used and the polished-Ti and porous-Ti substrates were placed at a distance of 5 mm from the electrodes within a glass vial (Figure 1). Plasma was generated at a voltage of 5 V and pulse width of 3 μs for 30 min. The mineralizing solution was stirred using a magnetic stirrer (Ion Stir 7d, Central Kagaku Co., Ltd., Tokyo, Japan) at a rotational speed of 240 rpm. The duration of the solution plasma treatment was 30 min.

When the solution plasma surface modification technique was applied to "Ti-AT" and on "Porous-Ti-AT", they were named "Ti-AT-SPT" and "Porous-Ti-AT-SPT", respectively.

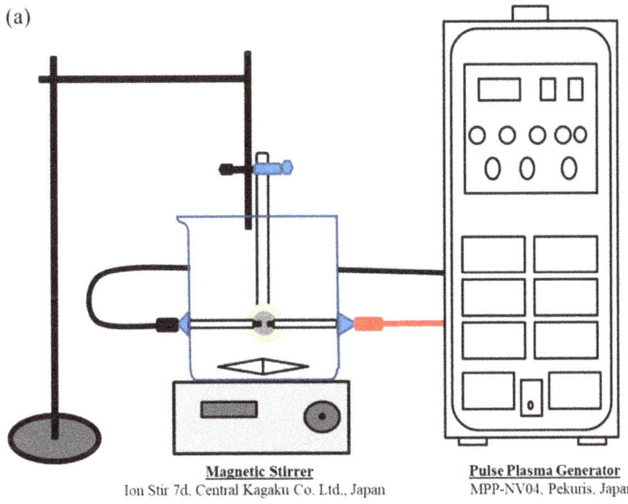

Figure 1. *Cont.*

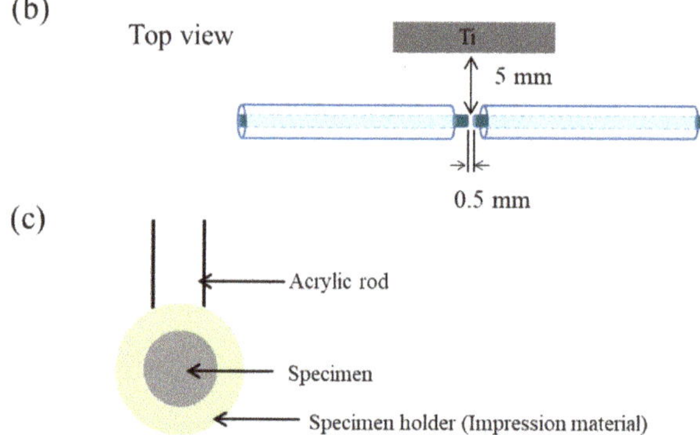

Figure 1. Apparatus, electrodes, and specimen employed for the solution plasma treatment. The specimen was placed at a 5-mm distance from the electrodes in a glass vial containing 150 mL of the mineralizing solution. Plasma was generated at 5 V and 3 μm pulse width for 30 min. (**a**) Pulse plasma generator, Magnetic stirrer, Holder stand; (**b**) Electrodes (2 electrode having 0.5 mm distance between them; (**c**) The specimen holder is made of Duplicone and acrylic rods help to hold the specimen holder.

2.4. Characterization of the HA Layer Formed on Various Substrates

The morphologic and crystallographic features of deposited films were examined by SEM and XRD. The contact angle of the water droplets (20 μL) on the titanium surfaces was measured by a contact angle analyzer after each surface treatment.

2.4.1. SEM Observation

The morphology of the titanium surface after SPT and the immersion treatments was examined using a scanning electron microscope (SSX-550, Shimadzu Corporation, Kyoto, Japan) with an acceleration voltage of 10 kV, after the samples were coated with gold.

2.4.2. X-ray Diffraction

The crystallographic features of the deposited films were examined using an X-ray diffractometer (JDX-3500, JEOL Ltd., Tokyo, Japan) with Cu Kα radiation ($\lambda = 0.1541$ nm) at room temperature. The 2θ range was 20–80°, and the XRD profile was recorded at step-scan intervals of 0.02° at a scanning speed of 8.0°/min at 40 kV and 150 mA.

2.4.3. Measurement of the Contact Angle of Water Droplets on the Surfaces

To evaluate the wettability of the titanium disks before and after surface treatment, the contact angle of the water droplets (20 μL) on the surfaces was measured using a contact angle analyzer (Phoenix Alpha, Seoul, Korea) at 0, 3, 7, and 14 days after each surface treatment.

2.5. Evaluation of Cytocompatibility

2.5.1. Cell Line

The MC3T3E1 cell line was used in this study to evaluate the cell attachment of the titanium samples before and after SPT.

2.5.2. Attachment of Cells on Differently Modified Titanium Surfaces after Four Hours of Incubation

There were 6 groups for each experiment, that is, 30-day aged ("Polished-Ti" and "Ti-AT-SPT"), 7-day aged ("Polished-Ti" and "Ti-AT-SPT"), fresh "Polished-Ti", and fresh "Ti-AT-SPT". Four specimens from each group were placed on a 24-well plate, disinfected with 1 mL of 70% ethanol for 10 min and washed with sterilized distilled water for 10 min. The number of MC3T3E1 cells was adjusted to be 5 × 10^4 cells/mL in a regular medium consisting of α-modified minimum essential medium (α-MEM; with L-glutamine and phenol red, Wako Pure Chemical Industries Ltd., Osaka, Japan) supplemented with 10% heat-inactivated fetal bovine serum. Adjusted cells were seeded on a sample with 1 mL per well. After 4 h of incubation under a humidified atmosphere of 5% CO_2 at 37 °C, the cells were detached by trypsin-EDTA (0.25% Trypsin/0.53 mM EDTA in Hanks Balanced Salt Solution without calcium or magnesium) and the number of cells was counted by a hemocytometer.

2.5.3. Statistical Analysis

Statistical analysis was performed using one-way analysis of variance (ANOVA), and the Games–Howell post hoc test with the significance level of $p < 0.05$.

3. Results

3.1. Variation of Solution Temperature in the Vicinity of the Sample with Time during SPT

Figure 2 shows the variation of the solution temperature with time during SPT measured by a thermocouple placed in the vicinity of the sample. The temperature increased rapidly and exceeded 60 °C within 3 min after starting SPT, and then increased slowly and reached an almost constant value of 80 °C at 30 min. After 30 min of SPT, crystals were precipitated due to heterogeneous nucleation both on the sample surface and in the solution.

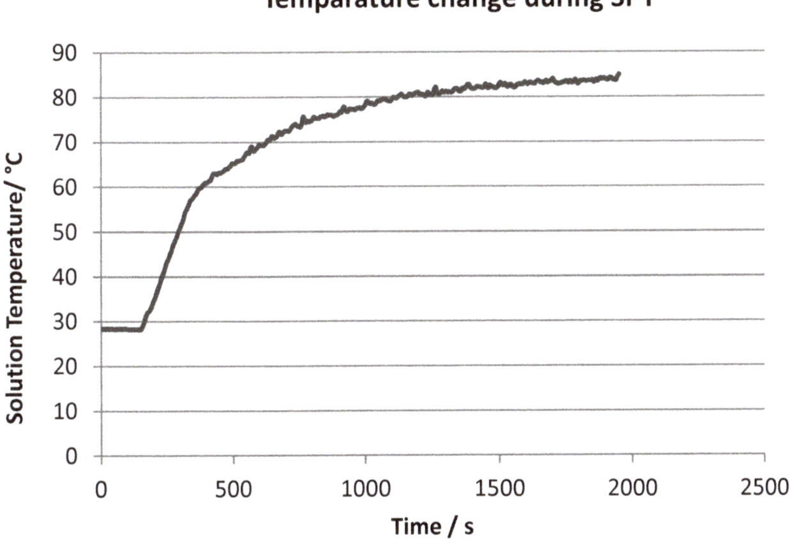

Figure 2. Variation of solution temperature in the vicinity of the sample with time, during solution plasma treatment (SPT). The temperature increased rapidly and exceeded 60 °C within 3 min after starting SPT, and then increased slowly and reached an almost constant value of 80 °C at 30 min.

3.2. Characterization of Titanium Surfaces Subjected to SPT and IT in the Calcium Phosphate Solution

3.2.1. SEM Observation

Figure 3 shows the SEM photographs of the polished titanium surface after SPT (Ti-SPT) (Figure 3a) and after 5 M NaOH treatment and subsequent SPT (Ti-AT-SPT) (Figure 3b). Crystals formed on both the Ti-SPT and the Ti-AT-SPT. Crystals with diameters between 5 and 20 μm precipitated sparsely and did not cover the whole surface of the Ti-SPT, while fine spherical crystals with a diameter of 5 μm precipitated and covered the whole surface of the Ti-AT-SPT.

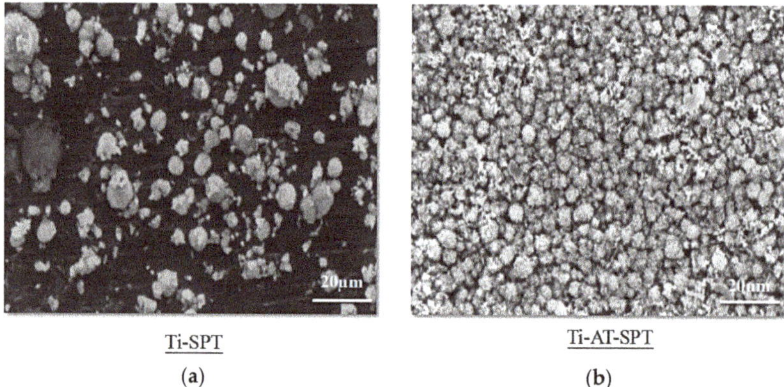

Figure 3. SEM images of HA crystals precipitated on Ti-SPT (**a**) and Ti-AT-SPT (**b**). The HA spherical particles formed on both the Ti-SPT and the Ti-AT-SPT. The spherical particles precipitated sparsely and did not cover the whole surface of the Ti-SPT, while the fine spherical crystals precipitated and covered the whole surface of the Ti-AT-SPT.

Figure 4 shows the SEM photographs of the porous titanium surface after 5 M NaOH treatment and subsequent SPT (Porous-Ti-AT-SPT). Fine spherical HA particles were uniformly precipitated over the entire surface (Figure 4a) including the areas recessed in the shape of the porous structure (Figure 4b) as well as the inner surface of the pores (Figure 4c).

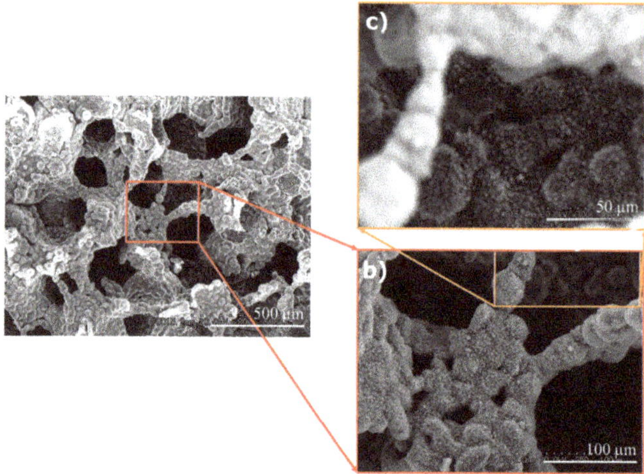

Figure 4. Scanning electron microscopy (SEM) images of Porous-Ti-AT-SPT. Fine spherical hydroxyapatite (HA) particles were uniformly precipitated over the entire surface (**a**), the areas recessed in the shape of the porous structure (**b**), and inner surface of the pores (**c**).

3.2.2. Analysis of Precipitated Particles on Ti-AT-SPT and Ti-AT-IT60° by XRD and EDX

Figure 5 shows the X-ray diffractogram obtained from Polished-Ti (Figure 5a), Ti-AT-SPT (Figure 5b), and synthetic HA powder (Figure 5c). The synthetic HA powder was obtained from the dried-out mineralizing solution which was subjected to SPT.

All of the diffraction peaks were assigned to HA or the titanium substrate under the deposited film. For Ti-AT-SPT (Figure 5b), the diffraction peak of the deposited HA crystals at 25.9° (2θ) that corresponded to the (002) lattice plane was relatively higher than the other diffraction peaks at around 32°, unlike that observed for the synthetic HA powder. This indicated that the spherical HA crystals that precipitated on Ti-AT-SPT were slightly oriented with the c-axis perpendicular to the titanium substrate. These results clearly demonstrated that 5 M NaOH treatment and subsequent SPT for 30 min in the calcium phosphate solution were effective in coating the entire surface of titanium with an HA film composed of fine crystals in a relatively short treatment time.

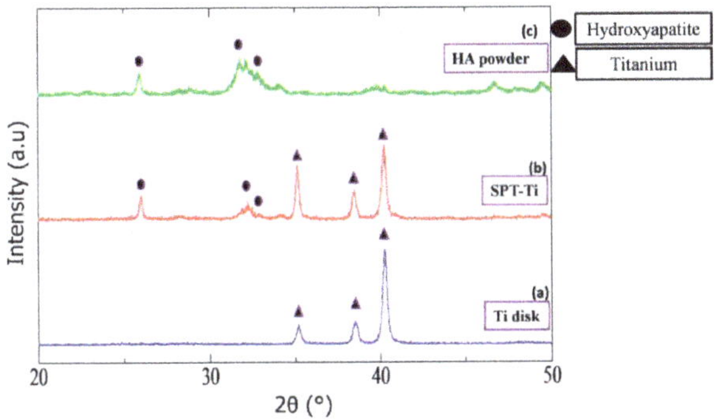

Figure 5. X-ray diffractogram obtained from (**a**) Polished-Ti, (**b**) Ti-AT-SPT, and (**c**) synthetic HA powder. All the diffraction peaks obtained from Ti-SPT were assigned to HA or the titanium substrate under the deposited film.

3.3. Crystals on the Titanium Surfaces after Immersion in the Calcium Phosphate Solution

Figure 6 shows the SEM photographs of the polished titanium after immersion in the calcium phosphate solution at 37 °C (Ti-IT37°) and after 5 M NaOH treatment and subsequent immersion in the calcium phosphate solution at 37 °C (Ti-AT-IT37°). No crystals were deposited on the Ti-IT37° for up to 7 days of immersion. In contrast, spherical particles consisting of plate-like HA crystals were deposited on part of the Ti-AT-IT37° after 1 day of immersion. These crystals grew with time and covered almost all of the surface of the titanium after immersion for 7 days. These results were consistent with those reported previously [5].

Figure 7 shows the SEM photographs of the polished titanium after immersion in the calcium phosphate solution at 60 °C (Ti-IT60°) and after 5 M NaOH treatment and subsequent immersion in the calcium phosphate solution at 60 °C (Ti-AT-IT60°). No spherical particle precipitation was observed on the Ti-IT60° after immersion in the solution for 30 min. Precipitated crystals between 5 and 40 μm in diameter were found on the Ti-IT 60° after immersion for 6 h, and these HA spherical particles grew into large size spherical particles between 40 and 100 μm in diameter after immersion for 24 h. On the Ti-AT-IT60°, small crystals precipitated after immersion for 30 min. The density of the crystals increased due to the growth of crystals with time. From the results shown in Figures 6 and 7, it was observed that the crystal formation and its growth rate were markedly enhanced with 5 M NaOH treatment as well as an increase in the temperature of the calcium phosphate solution.

Figure 6. SEM images of the Polished-Ti and Ti-AT after immersion in the meta stable calcium phosphate solution at 37 °C (Ti-IT37° and Ti-AT-IT37°). No HA spherical particles were deposited on the Ti-IT37° for up to 7 days of immersion. In contrast, spherical particles consisting of plate-like HA were deposited on the part of the Ti-AT-IT37° after immersion for 1 day.

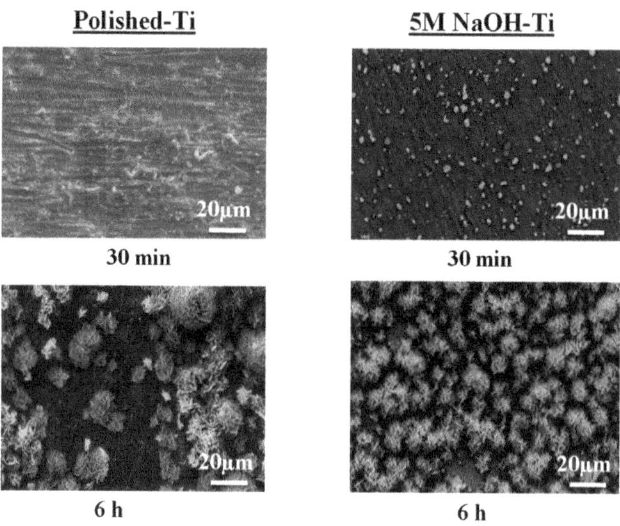

Figure 7. *Cont.*

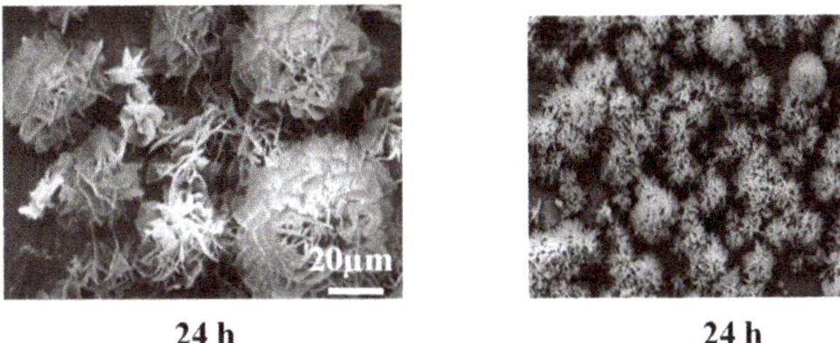

Figure 7. SEM images of the Polished-Ti and Ti-AT after immersion in the calcium phosphate solution at 60 °C (Ti-IT60° and Ti-AT-IT60°). There was no HA precipitation observed on the Ti-IT60° after immersion in the solution for 30 min. HA spherical particles were found on the Ti-IT60° after immersion for 6 h and they grew into large spherical particles after immersion for 24 h. On the Ti-AT-IT60°, small HA spherical particles precipitated after immersion for 30min. The spherical particles grew larger with time and covered the entire surface of a titanium after immersion for 24 h.

3.4. Change in the Contact Angle of Water Droplets on Polished-Ti and Ti-AT-SPT with Aging Time

Figure 8 shows the changes in the contact angle of 20 μL water droplets on Ti, Ti-AT, and Ti-AT-SPT with aging time. The contact angle value for Ti increased as the Ti disk aged, suggesting that the surface property changed from being hydrophilic to hydrophobic. This phenomenon is well known as the time-dependent degradation in biological capability or biological degradation, which arises from the absorption of hydrocarbon contaminants in air. In contrast, the contact angles for Ti-AT and Ti-AT-SPT were much lower than that for Ti throughout the aging period. The initial super hydrophilic surface remained almost unchanged up to 30 days of aging in air, although the tough hydrophilicity significantly decreased.

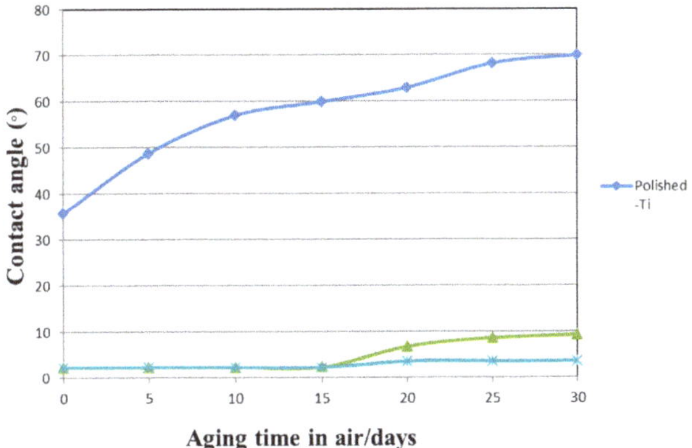

Figure 8. The changes in contact angle of water droplets on Polished-Ti, Ti-AT, and Ti-AT-SPT with aging time are shown. The contact angle for Ti-AT and Ti-AT-SPT was much lower than that for Ti throughout the aging period.

3.5. Bioactivity Analysis

Figure 9 shows the number of cells attached to the Polished-Ti and Ti-AT-SPT, which were subjected to aging in air for 0, 7, and 30 days, as measured with the hemocytometer. The number of MC3T3E1 cells attached to Ti-AT-SPT was approximately 50% more than the amount attached to Ti at each aging period, which suggested that the HA coating by solution plasma treatment significantly improved the initial cell attachment ($p < 0.05$). The number of cells that attached to the Polished-Ti with

30 days of aging in air was approximately 50% of that for the Polished-Ti without aging. In contrast, the number of attached cells on the Ti-AT-SPT after aging for 30 days remained above 85% of that for the SPT-Ti without aging.

Figure 10 shows the SEM photographs of the cells attached to Porous-Ti-AT-SPT after incubation for 4 h. The attached cells indicated by the arrows were located not only on the outer surface (Figure 10a) but also on the surface of the concavity (Figure 10b), which indicated that the induction of bone tissue on the entire porous surface was going to occur successfully and that bone tissue growth could be expected.

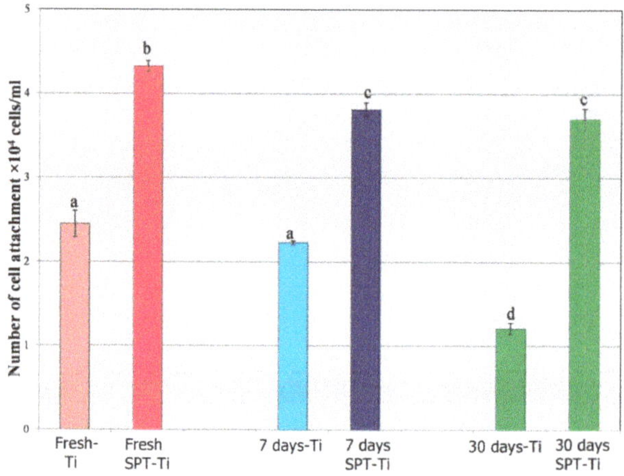

Figure 9. The number of attached cells on surfaces after 4 h incubation. The number of MC3T3E1 cells attached to the Polished-Ti and Ti-AT-SPT with 0, 7, and 30 days aging in air is shown in the SEM and graph. Different letters in the graph mean statistical difference at $p < 0.05$.

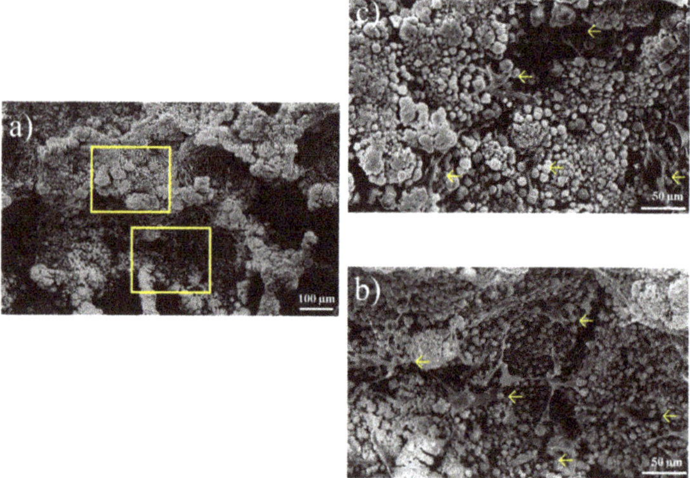

Figure 10. (**a**) Cell attached to porous Ti-AT-SPT after 4h incubation. (**b**) Cell attached at the outer surface of concavity in the porous structure. (**c**) Cell attached at the inner surface of concavity in the porous structure. SEM images of the cells attached to Porous Ti-AT-SPT after 4 h incubation. The attached cells indicated by arrows were located not only on the outer surface (upper yellow square represents c) but also on the surface of concavity (lower square represents b).

4. Discussion

4.1. Process of Rapid Formation of Thin and Uniform HA Film Consisting of Fine Spherical Particles on Ti-AT-SPT

With SPT in the calcium phosphate solution, HA spherical particles were precipitated on the Polished-Ti surface, but the size of the spherical particles was 5–20 μm in diameter. The density of the HA crystals was too low to form an HA film (Figure 3a). However, it was demonstrated that a thin and uniform HA film consisting of fine spherical HA particles with a diameter of 5 μm formed on Ti-AT-SPT (Figure 3b). This rapid formation of the uniform HA film was attributed to the synergistic effect obtained by combining an alkaline treatment with SPT in a calcium phosphate solution.

The driving force for the formation of the HA film from a supersaturated solution was the change in Gibbs free energy, ΔG, for transfer from a supersaturated solution to an equilibrium solution with HA crystals:

$$\Delta G = -RT \ln S \tag{1}$$

where R is the gas constant; T is the absolute temperature; and S is the degree of supersaturation, which is expressed as:

$$S = IP/K_{sp} \tag{2}$$

where IP is the ionic activity and K_{sp} is the thermodynamic solubility product. Since the solubility product of HA, K_{sp}, decreases with increasing temperature, it is apparent that an increase in the degree of supersaturation, S, together with a decrease in the K_{sp} value as a result of SPT was responsible for the precipitation of HA crystals on Polished-Ti. SPT for 30 min without alkaline treatment, however, was insufficient for the formation of a dense and uniform HA film. This probably arose from the fact that the SPT was not sufficient to lower the activation energy to enhance the formation of HA crystals (Figure 11a) and a few critical and supercritical HA nuclei were induced by SPT, as shown in Figure 12a.

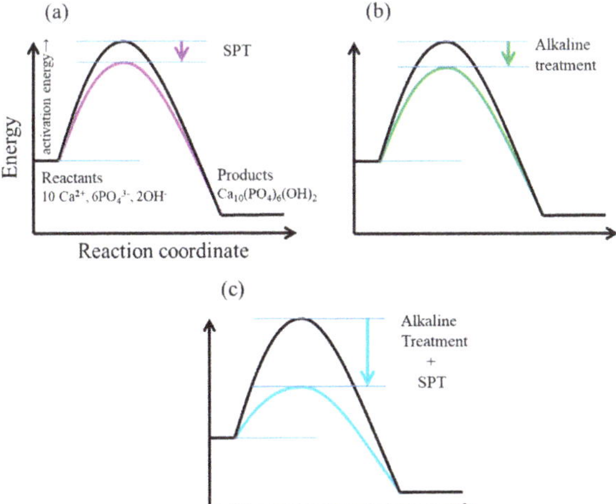

Figure 11. Energy profile in the hydroxyapatite formation. Activation energy of hydroxyapatite formation was lowered by (**a**) SPT and (**b**) alkaline treatment, and was notably lowered by (**c**) SPT and alkaline treatment and SPT.

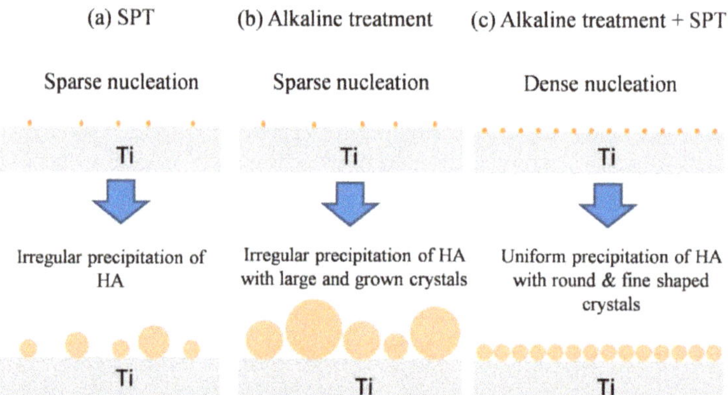

Figure 12. Schematic illustration of HA nucleation and HA crystal growth on the titanium surfaces. The appearance of particles formed on the Ti surface was quite different among (**a**) Ti-SPT, (**b**) Ti-AT-IT, and (**c**) Ti-AT-SPT.

With 5 M NaOH treatment, a sodium titanate hydrogel layer formed on the polished titanium [5]. It was also found that the hydrogel layer quickly released sodium ions with the uptake of calcium ions when alkaline-treated Ti was soaked in calcium phosphate solution. This ion exchange reaction was proposed to take place very quickly to maintain the electrical neutrality of the hydrogel and increased the calcium ion concentration at the hydrogel surface. Increase in *IP*, as a result of a higher calcium ion concentration, also increased the degree of supersaturation with respect to HA, which enhanced the apatite nucleation in a mineralizing solution [13]. In this case, the activation energy required for HA nucleation was also insufficiently lowered by the increased concentration of Ca ions (Figure 11b) to induce a few nuclei (Figure 12b). The HA crystals grew larger with an increase in soaking time in a calcium phosphate solution, and a thick HA film consisting of coarse HA spherical particles was formed after soaking for 1 day at 60 °C (Figure 11b).

With the combination of alkaline treatment and SPT, the degree of supersaturation, *S*, was markedly increased with both an increase in *IP* and a decrease in K_{sp}. The activation energy required for HA nucleation was probably lowered sufficiently (Figure 11c) to induce a large number of HA nuclei as shown in Figure 12c. As a result, a thin and uniform HA film consisting of fine spherical particles could be obtained (Figure 12c). As it is a uniform precipitation with fine and round HA particles, the height of the precipitation is equal to the height of HA crystal, which is 1 μm.

4.2. Chemical and Biological Properties of HA Film Formed by SPT

Previous studies have revealed that titanium undergoes a time-dependent degradation in biological capability due to the unavoidable contamination of the titanium surfaces by hydrocarbons in air [14]. Contact angle measurements of the water droplets demonstrated that the titanium surface changed from being hydrophilic to hydrophobic with an increase in aging time (Figure 8). This study demonstrated that the contact angle value for a HA-coated titanium disk (Ti-AT-SPT) was significantly lower than that for polished titanium throughout the aging periods. In addition to this, the initial super hydrophilicity of the HA-coated titanium surface by SPT was found to be maintained after 30-day aging in air (Figure 8). Unlike titanium, the HA-coated titanium did not show age-related impaired bioactivity, which is defined as biological aging [15]. The reason for the highly hydrophilic nature of HA-coated titanium can be attributed to the characteristic of HA crystals as HA is an ionic crystal with an abundance of ionic sites that attract water molecules.

The number of MC3T3E1 cells attached to Ti-AT-SPT was approximately 50% larger than that of Polished-Ti at each aging period, suggesting that the HA coating by solution plasma treatment significantly improved the initial cell attachment. This may be one of the reasons why HA-coated titanium implants possess better cytocompatibility and facilitate rapid bone formation due to their excellent osteoconductive property.

These results suggest that an osteoconductive porous-surfaced titanium implant can be developed with a HA coating by solution plasma treatment in a calcium phosphate solution in a short time.

4.3. HA Film Coating on Titanium by SPT to Develop an Osteoconductive Porous-Surfaced Dental Implant Body

A HA film coating was achieved when Ti-AT was soaked in calcium phosphate solution at 37 °C for seven days and at 60 °C for one day (Figures 6 and 7).

In a 37 °C immersion treatment, a HA film coating was achieved when polished-Ti and Ti-AT were soaked in calcium phosphate solution for seven days. The approximate size of the precipitated HA spherical particles on Ti-AT was 40–50 μm. In a 60 °C immersion treatment, an HA film coating was achieved when the Polished-Ti and Ti-AT were soaked in the calcium phosphate solution for 24 h. The approximate size of the precipitated HA spherical particles on Polished-Ti was more than 40 μm, and was 20 μm on Ti-AT.

These results suggest that the size of the HA spherical particles precipitated during the immersion treatments in the calcium phosphate solution was too large to coat the titanium implant with a porous surface structure. Furthermore, in the 37 °C immersion, there was less precipitation compared to the 60 °C immersion. For a dense precipitation of HA, the temperature must be 60 °C or more. Temperature of the mineralizing solution must be 37 °C during the initializing of the solution plasma treatment.

With 5 M NaOH treatment and subsequent SPT in the calcium phosphate solution, dense nucleation with critical size nuclei was successfully precipitated on the sample (Figure 12c) and a thin and dense HA layer consisting of fine spherical crystals was formed on the titanium disk surface after SPT for 30 min (Figure 3b).

The solution plasma surface modification for coating HA onto titanium proceeds in a wet process unlike other HA surface methods using a dry process. Therefore, the coating of a HA film consisting of fine crystals on the entire surface of titanium implant bodies with porous surface structures can be achieved. MC3T3E1 cells were observed not only on the outer surface, but also on the surface of the concavity (Figure 10), indicating that the induction of bone tissue on the entire porous surface would occur successfully and that bone tissue ingrowth could be expected. To enhance the growth of mineralized tissue into the pore spaces and to keep the vascular system interconnected by the pores for continued bone development, the porous structure of a titanium implant provides a unique biological bone anchorage to the titanium implants. Enhancement of bone/implant mechanical bonding can allow dentists to use dental implant bodies with a short length, which can greatly contribute to an increase in the ability of dental implants to be used for a wide variety of treatment options, especially for patients with low bone quality and with advanced maxillary/mandibular residual ridge resorption.

4.4. Features of the SPT for HA Film Coating

A HA film coating on titanium was achieved by combining a simple chemical treatment (5 M NaOH treatment) with SPT in a calcium phosphate solution. One of the striking features of this proposed method was that the rate of HA film formation during SPT was faster than that in most wet processes. For example, Kim et al. reported that a bone-like apatite layer formed on the 5 M NaOH treated titanium substrate when the substrate had been soaked in simulated body fluid (SBF) for more than one day [5]. In contrast, a uniform HA film was observed to form on the 5 M NaOH treated titanium by SPT for only 30 min (Figure 3b). As stated before, the thermal energy produced during SPT increased the degree of solution supersaturation with respect to HA, and thus greatly increased the growth rate of the HA crystals.

The substrate heating method in liquid with the application of a large current can be applied to electrically conducting materials such as metals and alloys. For SPT, the solution in contact with titanium is heated with the thermal energy generated by plasma, suggesting that this HA coating method can be applied to materials without electrical conductivity such as ceramics.

5. Conclusions

The present study demonstrated that it is possible to acquire a homogenous precipitation of hydroxyapatite over the entire surface of a smooth and porous titanium disk by the implementation of a solution plasma surface modification treatment in a mineralizing solution. The solution plasma-treated samples showed significantly better cyto-compatibility than the other specimens.

Author Contributions: Conceptualization K.E.; Methodology, K.E. and T.S.; Software, A.B.D. and M.R.H.; Validation, F.N.-T. and T.N.; Formal Analysis, A.B.D., T.N. and F.N.-T.; Investigation, A.B.D.; Resources, K.E.; Data Curation, A.B.D.; Writing-Original Draft Preparation, K.E. and A.B.D.; Writing-Review & Editing, T.S. and M.R.H.; Visualization, A.B.D. and M.R.H.; Supervision, K.E. and T.N.

Funding: This research received no external funding.

Acknowledgments: The authors would like to acknowledge the Division of Orthodontics and Dentofacial Orthopedics, School of Dentistry, Health Sciences University of Hokkaido for their technical support.

Conflicts of Interest: The authors declare no conflict of interest.

References

1. Dental Implants Facts and Figures. Available online: https://www.aaid.com/about/Press_Room/Dental_Implants_FAQ.html (accessed on 1 December 2018).
2. Herman, H. Plasma spray deposition processes. *MRS Bull.* **1988**, *12*, 60–67. [CrossRef]
3. Dasarathy, H.; Riley, C.; Coble, H.D. Analysis of apatite deposits on substrates. *J. Biomed. Mater. Res.* **1993**, *27*, 477–482. [CrossRef] [PubMed]
4. Ong, J.L.; Harris, L.A.; Lucas, L.C.; Lacefield, W.R.; Rigney, D. X-ray photoelectron spectroscopy characterization of ion-beam sputter-deposited calcium phosphate coatings. *J. Am. Ceram. Soc.* **1991**, *74*, 2301–2304. [CrossRef]
5. Kim, H.M.; Miyaji, F.; Kokubo, T.; Nishiguchi, S.; Nakamura, T. Graded surface structure of bioactive titanium prepared by chemical treatment. *J. Biomed. Mater. Res.* **1999**, *45*, 100–107. [CrossRef]
6. Ban, S.; Maruno, S.; Harada, A.; Hattori, M.; Narita, K.; Hasegawa, J. Effect of temperature on morphology of electrochemically deposit calcium phosphates. *Dent. Mater. J.* **1996**, *15*, 31–38. [CrossRef] [PubMed]
7. Ishikawa, K.; Miyamoto, Y.; Nagayoma, M.; Asaoka, K. Blast coating method: New method of coating titanium surface with hydroxyapatite at room temperature. *J. Biomed. Mater. Res.* **1997**, *38*, 129–134. [CrossRef]
8. Kuroda, K.; Ichino, R.; Okido, M.; Takai, O. Hydroxyapatite coation on titanium by thermal substrate method in aqueous solution. *J. Biomed. Mater. Res.* **2001**, *59*, 390–397. [CrossRef] [PubMed]
9. Tamura, M.; Endo, K.; Maida, T.; Ohno, H. HA film coating by the thermally induced liquid-phase deposition method for titanium implants. *Dent. Mater. J.* **2006**, *25*, 32–38. [CrossRef] [PubMed]
10. Saito, G.; Nakasugi, Y.; Akiyama, T. Generation of solution plasma over a large electrode surface area. *J. Appl. Phys.* **2015**, *118*, 23303. [CrossRef]
11. Kokubo, T.; Kim, H.M.; Kawashita, M.; Nakamura, T. Bioactive metals: Preparation and properties. *J. Mater. Sci. Mater. Med.* **2004**, *15*, 99–107. [CrossRef] [PubMed]
12. Saito, T.; Arsenault, A.L.; Yamauchi, M.; Kuboki, Y.; Crenshaw, M.A. Mineral Induction by immobilized phosphoproteins. *Bone* **1997**, *21*, 305–311. [CrossRef]
13. Nygren, H. Initial reactions of whole blood with hydrophilic and hydrophobictitanium surface. *Colloids Surf. B Biointerfaces* **1996**, *6*, 329–333. [CrossRef]

14. Jia, S.; Zhang, Y.; Ma, T.; Chen, H.; Lin, Y. Enhanced hydrophilicity and protein adsorption of titanium surface by sodium bicarbonate solution. *J. Nanomater.* **2015**, *2015*, 5. [CrossRef]
15. Att, W.; Hori, N.; Takeuchi, M.; Ouyang, J.; Yang, Y.; Anpo, M.; Ogawa, T. Time-dependent degradation of titanium osteoconductivity: An implication of biological aging of implant materials. *Biomaterials* **2009**, *30*, 5352–5363. [CrossRef] [PubMed]

Article

Electrophoretic Deposition as a New Bioactive Glass Coating Process for Orthodontic Stainless Steel

Kyotaro Kawaguchi [1], Masahiro Iijima [1,*], Kazuhiko Endo [2] and Itaru Mizoguchi [1]

[1] Division of Orthodontics and Dentofacial Orthopedics, School of Dentistry, Health Sciences University of Hokkaido, Hokkaido 061-0293, Japan; kawaguchi320@hoku-iryo-u.ac.jp (K.K.); mizo@hoku-iryo-u.ac.jp (I.M.)

[2] Division of Biomaterials and Bioengineering, School of Dentistry, Health Sciences University of Hokkaido, 1757 Kanazawa, Ishikari-Tobetsu, Hokkaido 061-0293, Japan; endo@hoku-iryo-u.ac.jp

* Correspondence: iijima@hoku-iryo-u.ac.jp; Tel.: +81-133-23-2977

Academic Editor: Saber AminYavari

Received: 6 October 2017; Accepted: 10 November 2017; Published: 13 November 2017

Abstract: This study investigated the surface modification of orthodontic stainless steel using electrophoretic deposition (EPD) of bioactive glass (BG). The BG coatings were characterized by spectrophotometry, scanning electron microscopy with energy dispersive X-ray spectrometry, and X-ray diffraction. The frictional properties were investigated using a progressive load scratch test. The remineralization ability of the etched dental enamel was studied according to the time-dependent mechanical properties of the enamel using a nano-indentation test. The EPD process using alternating current produced higher values in both reflectance and lightness. Additionally, the BG coating was thinner than that prepared using direct current, and was completely amorphous. All of the BG coatings displayed good interfacial adhesion, and Si and O were the major components. Most BG-coated specimens produced slightly higher frictional forces compared with non-coated specimens. The hardness and elastic modulus of etched enamel specimens immersed with most BG-coated specimens recovered significantly with increasing immersion time compared with the non-coated specimen, and significant acid-neutralization was observed for the BG-coated specimens. The surface modification technique using EPD and BG coating on orthodontic stainless steel may assist the development of new non-cytotoxic orthodontic metallic appliances having satisfactory appearance and remineralization ability.

Keywords: electrophoretic deposition; enamel remineralization; bioactive glass; spectrophotometry; nanoindentation

1. Introduction

Many orthodontic materials are formed from metals, which typically have superior mechanical properties compared with other materials. However, there are aesthetic issues with metal orthodontic materials [1]. More aesthetically attractive orthodontic materials are desirable, especially for adult patients. Aesthetic brackets made from ceramics and plastics have been widely used in clinical orthodontics [2,3]. Unfortunately, ceramic brackets have shortcomings stemming from their brittle nature, e.g., occasional fracture when tying the ligature and fracture from archwire forces, along with tooth wear during treatment and enamel fracture at debonding [1,4]. Plastic brackets also have deficiencies, such as a tendency to discolor, wear and creep due to their poor mechanical properties [1,5]. To overcome these issues, glass fiber-reinforced polymer wires have been investigated [6–9] but have yet to be used widely because of their brittleness and inability to withstand sufficient force [6–8]. Recently, coated archwires, including metal wires coated with polymers and rhodium-plated wires, have been developed [10–15]. These are preferred by many patients and orthodontists because of their improved aesthetic qualities. However, polymer-coated wire loses a significant amount of its coating layer when used in the areas of archwire engagement [11,12], which affects frictional properties and bacterial adhesion [13,15].

Acid-etching of enamel surfaces for bracket bonding procedures has been accepted in modern clinical orthodontics since the direct bonding of orthodontic brackets to enamel was introduced in the mid-1960s [16,17]. The enamel surface around bonded brackets etched with phosphoric acid is more susceptible to demineralization because the areas stagnate with plaque, making tooth-cleaning more difficult and limiting the efficacy of natural self-cleaning mechanisms. Additionally, the mechanical properties of the enamel surface region decreased after bracket bonding with the etch-and-rinse adhesive system [18], and irreversible alteration of the enamel might increase the risk of enamel micro-cracks forming during debonding procedures. Therefore, further demineralization of the enamel after bracket bonding should be prevented and, ideally, remineralization should be enhanced.

One reasonable way to enhance the remineralization of tooth surfaces is to increase the calcium or fluoride concentrations of oral fluids [19,20]. Various bioactive glass (BG) have been investigated since the first ones were reported by Hench et al. (1971) [21]. These studies have included their osteo-inductive behavior, ability to bond to both soft and hard tissues, the capacity of the glass to release ions (Ca, Na, Si), and the ability to form a hydroxyapatite layer [22–25]. More recently, attention has focused on their modification to further enhance osteogenic behavior, or on further compositional changes to introduce additional multifunctional properties such as antimicrobial activity [26]. If the surface of the metallic orthodontic materials can be modified with a BG, it may help to prevent the demineralization of tooth surfaces surrounding brackets and enhance remineralization after bracket debonding; these features are attractive in the clinical orthodontic setting.

Electrophoretic deposition (EPD) is a simple, rapid, and versatile coating technique, whereby colloidal particles suspended in a liquid medium migrate under the influence of an appropriate electric field and are deposited onto an electrode, leading to film formation and coatings with high microstructural homogeneity and tailored thickness [27,28]. Among the different techniques used for surface modification in the biomedical field, EPD is particularly attractive because it is does not require expensive equipment and can be used with colloidal BG particles to form complex-shaped orthodontic materials.

In this article, BG particles were deposited onto orthodontic stainless steel disks by an EPD process under various conditions, and the BG coating was characterized esthetically, morphologically, and compositionally using various methods. Additionally, the effects of the BG coating on the remineralization ability of etched dental enamel and frictional properties were investigated.

2. Materials and Methods

2.1. Materials

Mechanically polished stainless steel (SUS316) disk specimens (diameter: 14 mm; thickness: 2 mm; Nogata Denki Kogyo, Tokyo, Japan) were purchased and cleaned ultrasonically and subjected to the BG coating process. Non-coated specimens served as a control.

The BG (45.0% SiO_2 + 24.5% Na_2O + 24.5% CaO + 6.0% P_2O_5) was prepared by melting the raw materials in a platinum crucible at 1550 °C for 90 min using an electrically heated furnace (model SSFT-1520; Yamada Denki, Tokyo, Japan). The molten glass was rapidly quenched by malleating (rolling) between two stainless steel plates of 10-mm thickness. After cooling overnight, the glass was ground for 2 min in a vibrational rod mill (model TI-200; CMT Co., Fukushima, Japan) to yield a particle diameter of ca. 100 μm. The powders were further milled using a high-pressure gas-milling apparatus (Nano Jetmizer; Aishin Nano Technologies, Saitama, Japan) under a grinding pressure of 1.4 MPa to provide particles with a median diameter (D50) of 1.98 μm. Analysis of the BG by X-ray diffraction (XRD) confirmed its amorphous structure.

2.2. EPD Process

Suspensions containing 20 g/L of BG were prepared in 100 mL of distilled water by dispersing the particles via magnetic stirring and sonication (model UD-100; Tomy Seiko, Tokyo, Japan) for 600 s (The BG in distilled water was 0.2 g/mL). The stainless-steel specimens were encapsulated in a silicone impression material except for an exposed deposition area 14 mm in diameter. The EPD cell included two parallel stainless steel disk specimens as the deposition and counter electrodes; the distance between the electrodes was maintained at 3 mm. The coating was deposited using direct current (DC) or 1-kHz sine-wave alternating current (AC). Constant voltages (10 and 15 V) were applied and the deposition time was 10 min for all conditions. All measured pH values of the mixed suspension were pH 12.0 ± 0.1. After deposition, a coated specimen was gently removed from the suspension and dried at room temperature for 24 h before further characterization (Figure 1).

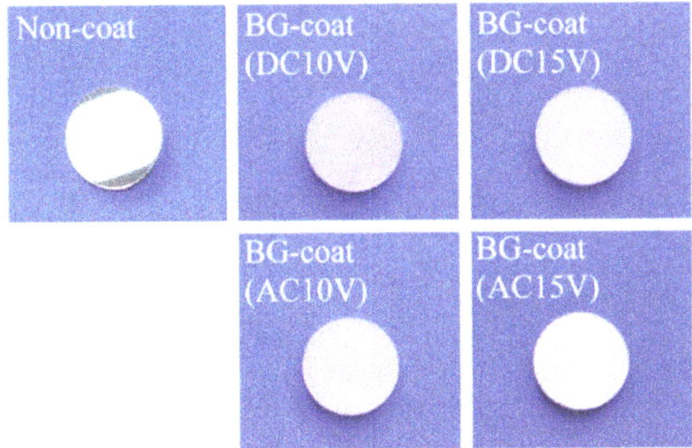

Figure 1. Photomicrographs of the non- and bioactive glass (BG)-coated specimens.

2.3. Color Measurements

The color of each specimen was measured using a spectrophotometer (model UV-2600; Shimadzu, Kyoto, Japan) with an integrated sphere (model ISR-2600 Plus; Shimadzu). Diffuse reflectance measurements were performed in the range of 350–800 nm in 1-nm steps. Color was measured according to the Commission International de L'Eclairage (CIE) $L^*a^*b^*$ color system [29] (n = 5), which has a lightness scale, L^*, and two opponent color axes, a^* and b^*. The redness and greenness are represented by the a^* values and the yellowness and blueness are represented by the b^* values.

2.4. Characterization of the Coatings

The surfaces of coated specimens were analyzed using XRD (model Rint-2500; Rigaku Corp., Tokyo, Japan) via a parallel-beam method using Cu–Kα radiation (40 kV; tube current: 100 mA). Representative specimens were analyzed over the 2θ range from 10 to 60° using a step size of 0.02° and a scan speed of 0.25°/min. The XRD patterns were obtained at 25 °C and analyzed qualitatively using PDXL2 software (Rigaku) based on the International Center for Diffraction Data (ICDD) database for phase identification and quantification.

To observe the coated layers and analyze their compositions on the cross-sectioned surface, specimens were encapsulated in an epoxy resin (Epofix; Struers, Copenhagen, Denmark) and cross-sectioned with a slow-speed, water-cooled diamond saw (Isomet; Buehler, Lake Bluff, IL, USA), then ground and polished using a series of silicon carbide abrasive papers and a final slurry of 0.05-μm alumina particles. All specimens were sputter-coated with pure gold for scanning electron microscopy (SEM) evaluation (model SSX-550; Shimadzu); the SEM was operated at 15 kV. The composition of a coated specimen was determined by energy-dispersive X-ray spectroscopy (EDS) at a working distance of 15 mm and a data acquisition time of 300 s.

2.5. Mechanical Properties of the Coatings

The external surfaces of the specimens were investigated with a nano-indentation apparatus (model ENT-1100a; Elionix, Tokyo, Japan). The specimens were fixed to the specimen stage with adhesive resin (Superbond Orthomite; Sun Medical, Shiga, Japan). Nano-indentation testing was carried out at 28 °C using a Berkovich indenter with a 10-mN peak load for ca. 1000 nm depth analysis (n = 10). Linear extrapolation methods (ISO Standard 14577) were used for the unloading curve between 95% and 70% of the maximum test force to calculate the elastic modulus [30–33]. The hardness and elastic modulus of the buccal enamel surfaces were calculated using the software bundled with the nano-indentation apparatus.

2.6. Frictional Properties Measured by the Progressive Load Scratch Test

A microtribometer (model CETR-UMT-2; Bruker, Billerica, MA, USA) was used to characterize the frictional properties by the progressive load scratch test. A diamond stylus having a 12.5-μm tip radius was moved 5 mm over a specimen surface with linearly increasing normal load (0.5 to 20 gf) at a constant speed of 0.016 mm/s, and the value of the friction coefficient was obtained (n = 5).

2.7. Acid-Neutralizing Ability

To estimate the acid-neutralizing ability, disk specimens were immersed in individual plastic vials containing 10 mL of acetic acid solution (pH = 4.5) at 37 °C. The time-dependent changes in the pH of the solutions were measured (model SI600; Sentron, Roden, The Netherlands) with a micro-pH electrode (model 9070-008; Sentron) over 24 h (n = 5).

2.8. Enamel Remineralization Ability and Changes in the Mechanical Properties

A total of 38 human non-carious premolars, obtained by extraction from patients undergoing orthodontic treatment, were cut with a slow-speed, water-cooled diamond saw (Isomet; Buehler, Lake Bluff, IL, USA) so that they were divided into mesial and distal halves; the sectioned specimens were then encapsulated in epoxy resin (Epofix; Struers, Copenhagen, Denmark). This in vitro study was approved by the ethics committee of the Health Sciences University of Hokkaido. After 24 h, the specimens were lightly ground with 600-grit sandpaper, and polished progressively using diamond suspensions with particle sizes of 3, 1, and 0.25 μm to obtain a surface suitable for nano-indentation. This polishing procedure removed approximately 200 μm of the tooth surface; a total of 75 polished-surface enamel specimens with an approximate area of 4 × 4 mm^2 were finally obtained. The specimens were divided into five groups of 15 specimens. Embedded human enamel specimens were etched with 35% phosphoric acid gel (Transbond XT Etching Gel; 3M Unitek, Monrovia, CA, USA) for 15 s, washed for 20 s, and dried in an air stream. Each etched enamel and disk specimen was fixed on a specimen stage and then immersed in a plastic vial containing 10 mL of artificial saliva at 37 °C for 3 months, with the solution changed weekly. Nano-indentation testing of the enamel surfaces was carried out at 28 °C (model ENT-1100a; Elionix) using two different loads (10 and 100 mN), before and after etching and during immersion periods. The hardness and elastic modulus of the buccal enamel surfaces were calculated.

2.9. Cytocompatibility

Mouse L929 fibroblast cells were seeded at a density of 5000 cells/cm^2 in 96-well plates and incubated in MEMα (Wako Pure Chemical, Osaka, Japan) containing 5% fetal bovine serum (ICN Biomedicals, Irvine, CA, USA) for 36 h at 37 °C with 5% CO_2. Each disk specimen was immersed into 10 mL of culture solution for 24 h and then 100 μL of the solution was added to each 96-well plate, followed by 24 h incubation. Finally, the absorbance of each well at 450 nm was recorded using a microplate reader (model Infinite F200; Männedorf, Switzerland) (n = 10).

2.10. Statistical Analyses

Statistical analyses were performed using the PASW Statistics software (ver. 18.0 J for Windows; IBM, Armonk, NY, USA). The mean values with standard deviations obtained for the various experiments in this study were compared using one-way analysis of variance (ANOVA) followed by Tukey's test. For all statistical tests, significance was predetermined at $p < 0.05$.

3. Results

3.1. Color Measurements

Figure 2 shows the changes in representative diffuse reflectance curves for the specimens. The lightness (L^*) and two opponent color (a^*, b^*) values are summarized in Table 1. The reflectance values (%) increased with increasing wavelength for all specimens. Non-coated specimens showed significantly higher reflectance values in the 350–800 nm range and L^*, due to the polished bright surface acting as a mirror. Among the BG-coated specimens, the one prepared using AC 15 V showed the highest values for both reflectance in the 350–800 nm range and L^*, followed by specimens coated at DC 15 V, AC 10 V, and DC 10 V. The specimens coated at higher voltage (15 V) showed significantly higher reflectance and L^* values than those coated at lower voltage (10 V). Similar a^* and b^* values were obtained for most BG-coated specimens.

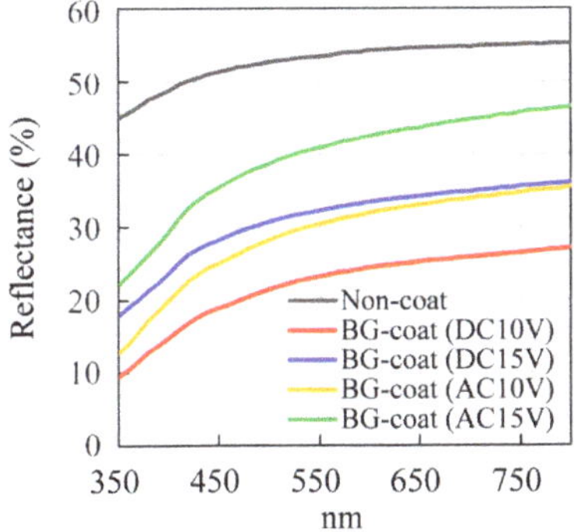

Figure 2. Diffuse reflectance curves of the various specimens.

Table 1. Commission International de L'clairage (CJE) lightness (L^*) and color (a^*, b^*) values for the specimens.

Coordinate	Non-Coat		BG-Coat (DC 10 V)		BG-Coat (DC 15 V)		BG-Coat (AC 10 V)		BG-Coat (AC 15 V)		p Value
	mean	S.D.	mean	S.D.	mean	S.D.	mean	S.D.	mean	S.D.	
L^*	78.02 [a]	0.14	54.88 [a]	0.34	63.79 [c]	0.14	61.83 [d]	0.36	70.42 [e]	0.71	0.000
a^*	−0.12 [a]	0.03	−0.50 [b]	0.05	−0.48 [b]	0.10	−0.48 [b]	0.02	−0.41 [b]	0.11	0.000
b^*	2.31 [a]	0.10	7.89 [b]	0.30	6.12 [c]	0.65	8.12 [b]	0.10	6.70 [c]	0.26	0.000

Notes: Values are mean and standard deviation (S.D.), $n = 7$. One-way ANOVA followed by Tukey–Kramer multiple range test. [a–e] Idential letters indicate that mean values were not significantly different ($p < 0.05$).

3.2. Crystal Structures, Morphological Features, and Compositions of the Coating Layers

Figure 3 displays representative XRD spectra of non- and BG-coated specimens. Weak broad feature at around 32° was obtained for the BG-coated specimen (AC 15V) because of the amorphous structure. On the other hand, two peaks at $2\theta = 43.5$ and 51.0° associated with the austenite (γ-Fe) phase (ICDD PDF 01-071-4649) were observed for the BG-coated specimen (DC 15V), while the broad feature at around 32° was observed.

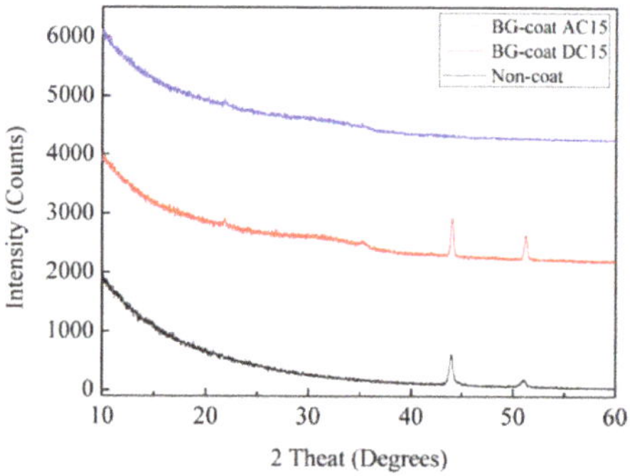

Figure 3. Representative X-ray diffraction (XRD) spectra of the non- and BG-coated specimens.

Figure 4 shows SEM photomicrographs and qualitative compositional maps obtained by EDS of BG-coated cross-sectional specimens. The thickness of the BG coating layers formed on the disk surfaces was ca. 1.0–4.0 μm and the specimens coated at higher voltage (15 V) tended to have thicker BG coating layers than those coated at lower voltage (10 V). Additionally, the thickness of the BG coating layers were similar in both specimens coated with AC and DC. Good interfacial adhesion was observed between all BG coating layers and the bulk materials. Si and O, the major components of BG, were enriched in all of the coating layers.

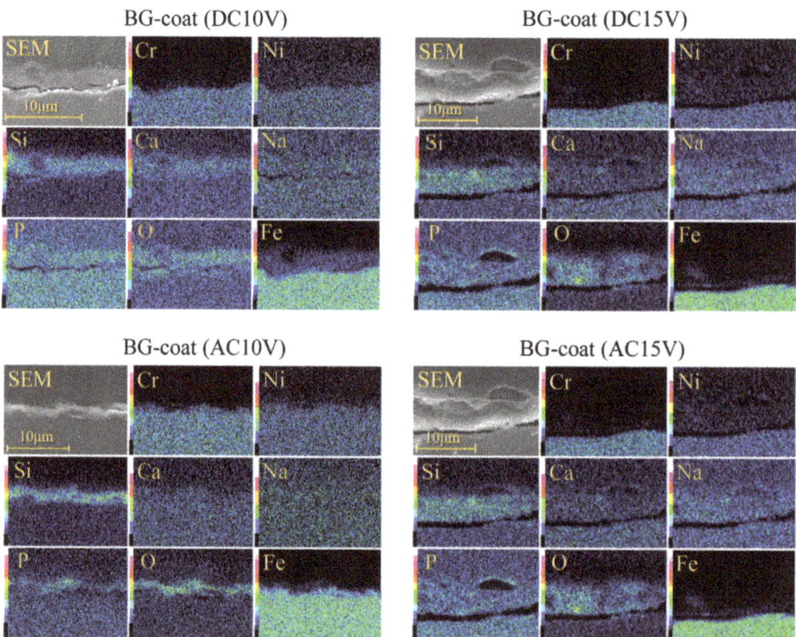

Figure 4. Scanning electron microscopy (SEM) photomicrographs and qualitative compositional maps obtained by energy-dispersive X-ray spectroscopy (EDS) of cross-sections of the BG-coated specimens.

3.3. *Evaluation of the Mechanical Properties of the Coating Layers by Nanoindentation*

Table 2 summarizes the mechanical properties of surface regions for non- and BG-coated specimens. The hardness and elastic modulus of the BG layers were significantly lower than those of the non-coated specimen. The specimens coated at higher voltage (15 V) showed significantly higher the hardness and elastic modulus of the BG layers than those coated at lower voltage (10 V). The elastic modulus of the BG layers formed by AC were significantly higher than those for the BG layers formed by DC.

Table 2. Mechanical properties of the non- and bioactive glass (BG)-coated specimens.

Mechanical Properties	Non-Coat		BG-Coat (DC 10 V)		BG-Coat (DC 15 V)		BG-Coat (AC 10 V)		BG-Coat (AC 15 V)		*p* Value
	mean	S.D.	mean	S.D.	mean	S.D.	mean	S.D.	mean	S.D.	
Hardness (GPa)	6.11 ^{a}	0.22	0.49 ^{a}	0.10	1.99 ^{c}	0.52	0.85 ^{b}	0.48	1.98 ^{c}	0.73	0.000
Elastic modulus (GPa)	192.46 ^{a}	4.96	70.47 ^{b}	23.22	109.12 ^{c}	23.39	84.32 ^{bc}	15.28	128.5 ^{d}	27.65	0.000

Notes: Values are mean and standard deviation (S.D.), $n = 7$; One-way ANO VA followed by Tukey–Kramer multiple range test. Identical letters indicate that mean values were not significantly different ($p < 0.05$).

3.4. *Frictional Properties Measured by the Progressive Load Scratch Test*

Figure 5 shows the change in representative frictional force obtained by the progressive load scratch test. The BG-coated specimens (DC 15 V) showed significantly higher frictional force than the other specimens when measured over 5 mm. Over approximately the first 4 mm of the measurement distance, the other BG-coated specimens displayed slightly higher frictional force than the non-coated specimen.

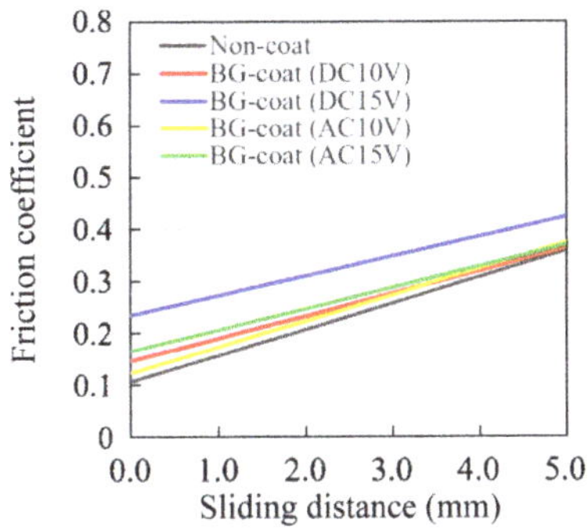

Figure 5. Frictional forces obtained by the progressive load scratch test.

3.5. *Analysis of Acid-Neutralizing Ability*

Figure 6 shows the time-dependent changes in the pH of the solutions as an indication of the acid-neutralizing ability. Immersion of BG-coated specimens in acetic acid solution caused an increase in the pH, and the acid-neutralizing ability increased with increasing output voltage.

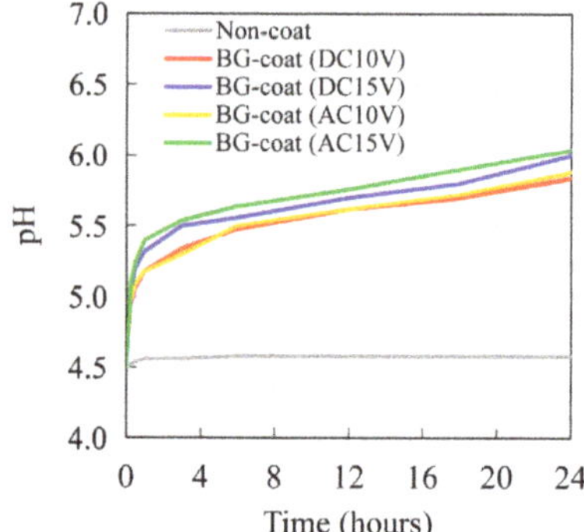

Figure 6. Time-dependent changes in the pH of the solutions as an indication of acid neutralizing ability.

3.6. Enamel Remineralization Ability in Artificial Saliva and Mechanical Property Changes Determined by Nanoindentation

Figures 7 and 8 show the mean hardness and elastic modulus values of the enamel surface before and after etching and during the 3 months of immersion. There was no significant difference in the hardness or elastic modulus between the groups before and immediately after etching. Phosphoric acid etching markedly decreased the hardness and elastic modulus of the enamel surfaces. The hardness and elastic modulus of the enamels of BG-coated specimens increased gradually with increasing immersion time. The recovery of the mechanical properties of a specimen immersed with a non-coated specimen was unremarkable. However, the hardness and elastic modulus of etched enamel specimens immersed with most BG-coated specimens recovered significantly compared with a non-coated specimen. Similar behavior was observed for both load conditions (10 and 100 mN).

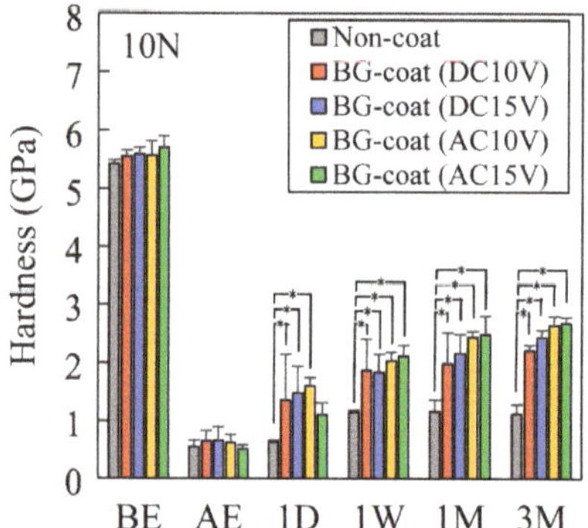

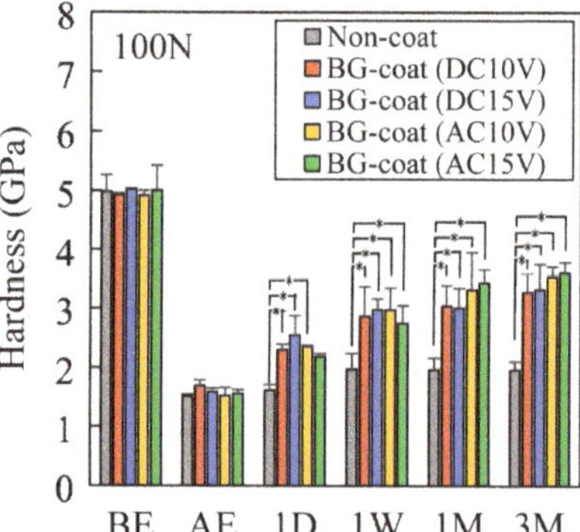

Figure 7. Mean hardness values of the enamel surface before and after etching, and during the 3-month immersion period. BE, before etching; AE, after etching; 1D, 1-day immersion; 1W, 1-week immersion; 1M, 1-month immersion; 3M, 3-month immersion.

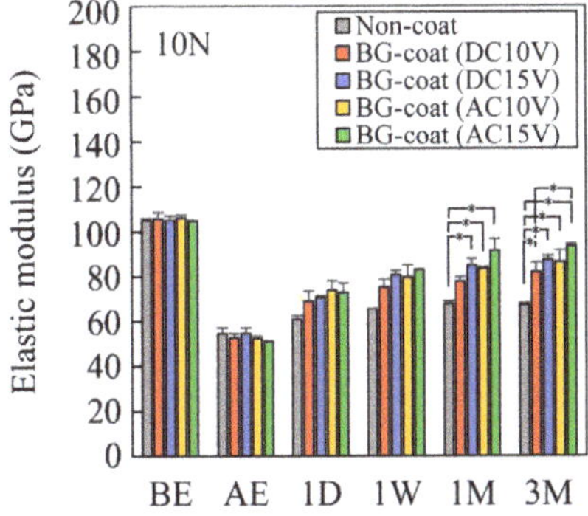

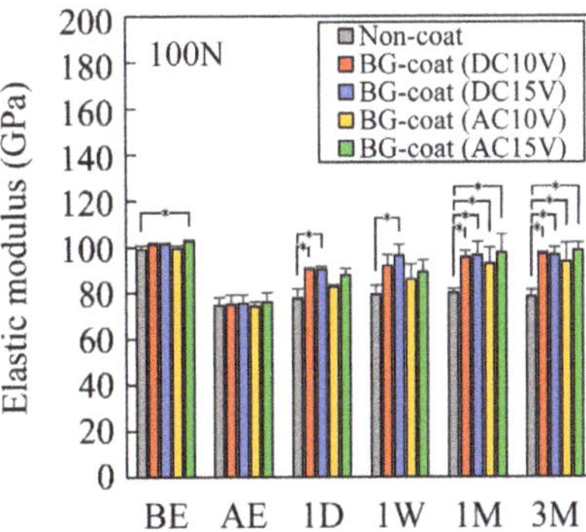

Figure 8. Mean elastic modulus values of the enamel surface before and after etching, and during the 3-month immersion period. BE, before etching; AE, after etching; 1D, 1-day immersion; 1W, 1-week immersion; 1M, 1-month immersion; 3M, 3-month immersion.

3.7. Cytocompatibility Assays

Figure 9 shows the absorbance at 450 nm as a function of the L929 fibroblast cell number at 36 h. No significant difference was noted in the mean fibroblast cell growth for 36 h, which indicated that the BG coatings were not cytotoxic.

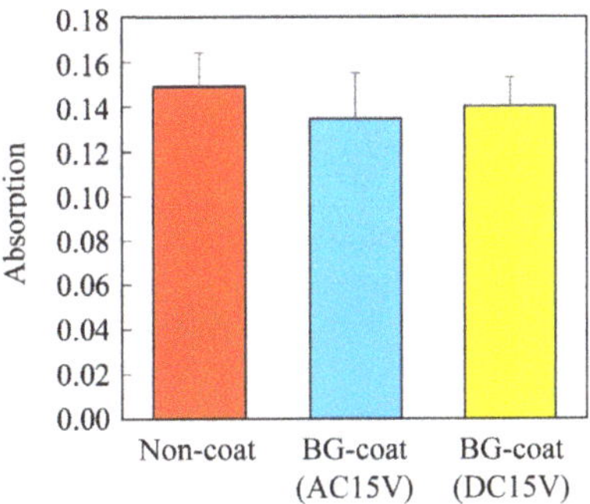

Figure 9. Absorbance at 450 nrn as a function of the L929 fibroblast cell number at 36 h.

4. Discussion

In the present study, thin BG coating layers with a milky-white appearance were formed on mechanically polished stainless steel specimens using an EPD process with a BG suspension and this is the first study that has investigate the esthetic performance of the BG coating. Quantitative color measurements showed that the EPD process using AC at 15 V produced higher values for both the reflectance (%) in the range of 350–800 nm and L^* (mean value: 70.42). The range of L^* values measured for the BG coating layers in the present study (54.88 to 70.42) was greater than that (36.2 to 50.3) reported for ceramic and plastic brackets [34], although the color values (a^*, −0.50 to −0.41; b^*, 6.12 to 8.12) for the BG coating layers were similar with published values (a^*, −1.3 to 3.8; b^*, −2.9 to 11.2) [34]. The color for orthodontic appliances, such as brackets and archwires, should ideally match that of natural teeth, although natural tooth color varies according to race, gender, and age [35]. A previous study measured the CIE $L^*a^*b^*$ color values for the Vita Lumin Vacuum shade guide (A3.5, B1, B3, C4), which is the color selection scale most widely used in dentistry; the values ranged from 43.2 to 61.4 for L^*, from −1.6 to 6.8 for a^*, and from 13.2 to 28.8 for b^*. The L^* and a^* values from the present study are fairly similar, although our mean b^* value was smaller than that of the Vita Lumin Vacuum shade guide. Thus, the BG coatings formed by the EPD process in the present study likely produced a clinically acceptable color appearance, but further research is warranted to fine-tune the color.

SEM observations showed that the thickness of the BG coating layers were similar in both specimens coated with AC and DC. However, surface-sensitive parallel-beam XRD analysis revealed that the peaks at $2\theta = 43.5°$ and $51.0°$ associated with the austenite (γ-Fe) phase of the bulk substrate were observed for the BG-coated specimens when DC (DC 15 V) was used. This suggested that the BG coating layer that formed on the specimen using DC had a too-low density. On the other hand, the XRD pattern obtained for thin BG coating layer formed on the specimen using AC (AC 15 V) indicated a completely amorphous phase, even for thinner BG coating layers, which suggested that thin BG coating layers formed using AC had better quality with high density. This difference in crystallinity is because undesired electrolysis of water at the electrodes occurred with the DC, which entrapped the coating and degraded the coating quality [36]. This was partly confirmed by the nano-indentation test results, which found that the elastic modulus of the BG layers formed by AC were significantly higher than those of the BG layers formed using DC.

The frictional force between the bracket and the archwire during tooth movement is a primary issue in orthodontics. If the frictional force can be decreased, then the efficiency of tooth movement can be improved [37,38]. The frictional properties are attributed to multiple factors, such as surface roughness, hardness, elastic modulus, and the cross-sectional dimensions of the orthodontic wires and brackets [37,38]. A recent study reported that commercially available esthetic coating wire influences the frictional properties [15] and the esthetic polymer coating may increase the frictional resistance due to increased wire-binding at the edge of the bracket. The progressive load scratch test used in the present study showed that the BG-coated specimens (DC 15 V) displayed significantly higher frictional forces than the other BG-coated specimens (DC 10 V; AC 10 and 15 V), although the other BG-coated specimens produced slightly higher frictional forces compared with the non-coated stainless steel specimen. The stainless steel wire alloy generally generates lower levels of frictional resistance than coated wires [15], and the BG-coated specimens, in some conditions, displayed frictional performance that was similar to that of the non-coated stainless steel specimens. Thus, the BG coating can likely provide acceptable clinical frictional characteristics. Evaluation of specimens with thin, high elastic modulus coatings are required to fully explore this aspect.

The results of this study showed that BG-coated specimens had significant acid-neutralizing capability due to their ability to release various ions. This suggests that the BG layer may be able to mitigate enamel demineralization [39,40]. Additionally, this in vitro study demonstrated that remineralization of etched enamel was accelerated for the BG-coated specimens, which is the most important finding of the present study. Nano-indentation testing with a 100-mN load showed that the hardness recovered by 49%–60%, and elastic modulus by 77%–84%, after 3 months of immersion. In comparison, the recovery of the mechanical properties of the etched enamel surface of non-coated specimens was unremarkable: The hardness recovered by 13% and the elastic modulus by 15%, although artificial saliva contains the inorganic ions necessary for remineralization [41]. A similar trend was observed for the mechanical properties of the top surface regions measured by nano-indentation testing with a 10-mN load (ca. 1000-nm depth analysis).

BG with an amorphous structure can release multiple ions into the oral environment [22–25], which may help to prevent the demineralization of tooth surfaces surrounding brackets and enhance remineralization after bracket debonding. In the present study, XRD analysis confirmed that the EPD coating process had little influence on the crystal structure of the BG coatings, and these enhanced enamel remineralization. However, the BG coatings slightly increased the frictional force. The BG coating formed using AC had acceptable quality and a completely amorphous structure, favorable esthetic character and mechanical properties. Thus, AC may be more suitable for the EPD coating process, although there was no significant difference between the BG-coated specimens formed by DC and AC in terms of their enamel remineralization ability. EPD as a new BG coating process in the present study can possibly produce acid-neutralizing and remineralizing capabilities of the enamel surfaces around orthodontic appliances. Further research to optimize the EPD coating conditions is warranted.

5. Conclusions

The surface modification technique using EPD and BG for orthodontic stainless steel offers the possibility of developing new orthodontic metallic appliances with satisfactory esthetic appearance and remineralization ability, without being cytotoxic.

Acknowledgments: The authors thank Masataka Fukuchi at Yokohama City Industrial Technical Support Center for his expert technical assistance with the progressive load scratch test. The authors also thank Kazuki Sobue and Takashi Murakami at Shimadzu Corporation for lending their expertise on the application of color measurement. This study was partially supported by a Grant-in-Aid Scientific Research from the Ministry of Education, Culture, Sports, Science and Technology, Japan (No. 25463195).

Author Contributions: Masahiro Iijima conceived and designed the experiments; Kyotaro Kawaguchi and Masahiro Iijima performed the experiments; Masahiro Iijima, Kazuhiko Endo, and Itaru Mizoguchi wrote the paper.

Conflicts of Interest: The authors declare no conflict of interest.

References

1. Iijima, M.; Zinelis, S.; Papageorgiou, S.N.; Brantley, W.; Eliades, T. Orthodontic brackets. In *Orthodontic Application of Biomaterials*; Eliades, T., Brantley, W., Eds.; Woodhead Publishing: Sawston, UK, 2017; pp. 75–96.
2. Iijima, M.; Muguruma, T.; Brantley, W.; Choe, H.C.; Nakagaki, S.; Alapati, S.B.; Mizoguchi, I. Effect of coating on properties of esthetic orthodontic nickel-titanium wires. *Angle Orthod.* **2012**, *82*, 319–325. [CrossRef] [PubMed]
3. Lopes Filho, H.; Maia, L.E.; Araújo, M.V.; Ruellas, A.C. Influence of optical properties of esthetic brackets (color, translucence, and fluorescence) on visual perception. *Am. J. Orthod. Dent. Orthop.* **2012**, *141*, 460–467. [CrossRef] [PubMed]
4. Karamouzos, A.; Athanasiou, A.E.; Papadopoulos, M.A. Clinical characteristics and properties of ceramic brackets: A comprehensive review. *Am. J. Orthod. Dent. Orthop.* **1997**, *112*, 34–40. [CrossRef]
5. Alrejaye, N.; Pober, R.; Giordano, I.R. Torsional strength of computer-aided design/computer-aided manufacturing-fabricated esthetic orthodontic brackets. *Angle Orthod.* **2017**, *87*, 125–130. [CrossRef] [PubMed]
6. Imai, T.; Yamagata, S.; Watari, F.; Kobayashi, M.; Nagayama, K.; Toyoizumi, H.; Uga, M.; Nakamura, S. Temperature-dependence of the mechanical properties of FRP orthodontic wire. *Dent. Mater. J.* **1999**, *18*, 167–175. [CrossRef] [PubMed]
7. Zufall, S.W.; Kusy, R.P. Sliding mechanics of coated composite wires and the development of an engineering model for binding. *Angle Orthodont.* **2000**, *70*, 34–47. [PubMed]
8. Burstone, C.J.; Liebler, S.A.H.; Goldberg, A.J. Polyphenylene polymers as esthetic orthodontic archwires. *Am. J. Orthod. Dent. Orthop.* **2011**, *139*, e391–e398. [CrossRef] [PubMed]
9. Tanimoto, Y.; Inami, T.; Yamaguchi, M.; Nishiyama, N.; Kasai, K. Preparation, mechanical, and in vitro properties of glass fiber-reinforced polycarbonate composites for orthodontic application. *J. Biomed. Mater. Res. B* **2015**, *103*, 743–750. [CrossRef] [PubMed]
10. Da Silva, D.L.; Mattos, C.T.; Simão, R.A.; De Oliveira Ruellas, A.C. Coating stability and surface characteristics of esthetic orthodontic coated archwires. *Angle Orthod.* **2013**, *83*, 994–1001. [CrossRef] [PubMed]
11. Da Silva, D.L.; Mattos, C.T.; Sant'Anna, E.F.; Ruellas, A.C.; Elias, C.N. Cross-section dimensions and mechanical properties of esthetic orthodontic coated archwires. *Am. J. Orthod. Dent. Orthop.* **2013**, *143*, S85–S91. [CrossRef] [PubMed]
12. Bradley, T.G.; Berzins, D.W.; Valeri, N.; Pruszynski, J.; Eliades, T.; Katsaros, C. An investigation into the mechanical and aesthetic properties of new generation coated nickel-titanium wires in the as-received state and after clinical use. *Eur. J. Orthod.* **2014**, *36*, 290–296. [CrossRef] [PubMed]

13. Kim, I.H.; Park, H.S.; Kim, Y.K.; Kim, K.H.; Kwon, T.Y. Comparative short-term in vitro analysis of mutans streptococci adhesion on esthetic, nickel-titanium, and stainless-steel arch wires. *Angle Orthod.* **2014**, *84*, 680–686. [CrossRef] [PubMed]
14. Rudge, P.; Sherriff, M.; Bister, D. A comparison of roughness parameters and friction coefficients of aesthetic archwires. *Eur. J. Orthod.* **2015**, 49–55. [CrossRef] [PubMed]
15. Muguruma, T.; Iijima, M.; Yuasa, T.; Kawaguchi, K.; Mizoguchi, I. Characterization of the coatings covering esthetic orthodontic archwires and their influence on the bending and frictional properties. *Angle Orthod.* **2017**, *87*, 610–617. [CrossRef] [PubMed]
16. Newman, G.V. Epoxy adhesives for orthodontic attachments: progress report. *Am. J. Orthod. Dent. Orthop.* **1965**, *51*, 901–912. [CrossRef]
17. Eliades, T. Orthodontic materials research and applications: Part 1. Current status and projected future developments in bonding and adhesives. *Am. J. Orthod. Dent. Orthop.* **2006**, *130*, 445–451. [CrossRef] [PubMed]
18. Iijima, M.; Muguruma, T.; Brantley, W.A.; Ito, S.; Yuasa, T.; Saito, T.; Mizoguchi, I. Effect of bracket bonding on nanomechanical properties of enamel. *Am. J. Orthod. Dent. Orthop.* **2010**, *138*, 735–740. [CrossRef] [PubMed]
19. Feathestone, J.D. Remineralization, the natural caries repair process—The need for new approaches. *Adv. Dent. Res.* **2009**, *21*, 4–7. [CrossRef] [PubMed]
20. Tschoppe, P.; Zandim, D.L.; Martus, P.; Kielbassa, A.M. Enamel and dentin remineralization by nano-hydroxyapatite toothpaste. *J. Dent.* **2011**, *39*, 430–437. [CrossRef] [PubMed]
21. Hench, L.L.; Hench, L.L.; Splinter, R.J.; Allen, W.C.; Greenlee, T.K. Bonding mechanisms at the interface of ceramic prosthetic materials. *J. Biomed. Mater. Res.* **1971**, *5*, 117–141. [CrossRef]
22. Kokubo, T. Apatite formation on surface of ceramics, metals and polymers in body environment. *Acta Mater.* **1998**, *46*, 2519–2527. [CrossRef]
23. Hoppe, A.; Güldal, N.S.; Boccaccini, A.R. A review of the biological response to ionic dissolution products from bioactive glasses and glass-ceramics. *Biomaterials* **2011**, *32*, 2757–2774. [CrossRef] [PubMed]
24. Fredholm, Y.C.; Karpukhina, N.; Brauer, D.S.; Jones, J.R.; Law, R.V.; Hill, R.G. Influence of strontium for calcium substitution in bioactive glasses on degradation, ion release and apatite formation. *J. R. Soc. Interface* **2012**, *9*, 880–889. [CrossRef] [PubMed]
25. Liu, J.; Rawlinson, S.C.; Hill, R.G.; Fortune, F. Fluoride incorporation in high phosphate containing bioactive glasses and in vitro osteogenic, angiogenic and antibacterial effects. *Dent. Mater.* **2016**, *32*, 412–422. [CrossRef] [PubMed]
26. Fernandes, J.S.; Gentile, P.; Pires, R.A.; Reis, R.L.; Hatton, P.V. Multifunctional bioactive glass and glass-ceramic biomaterials with antibacterial properties for repair and regeneration of bone tissue. *Acta Biomater.* **2017**, *59*, 2–11. [CrossRef] [PubMed]
27. Besra, L.; Liu, M. A review on fundamentals and applications of electrophoretic deposition (EPD). *Prog. Mater. Sci.* **2007**, *52*, 1–61. [CrossRef]
28. Chen, Q.; Garcia, R.P.; Munoz, J.; Pérez de Larraya, U.; Garmendia, N.; Yao, Q.; Boccaccini, A.R. Cellulose nanocrystals—Bioactive glass hybrid coating as bone substitutes by electrophoretic co-deposition: In situ control of mineralization of bioactive glass and enhancement of osteoblastic performance. *ACS Appl. Mater. Interfaces* **2015**, *7*, 24715–24725. [CrossRef] [PubMed]
29. Commission International de L'Eclairage. *Colormetry—Technical Report CIE Pub. No. 15*, 3rd ed.; Bureau Central de la CIE: Vienna, Austria, 2004.
30. *ISO 14577-1 Metallic Materials—Instrumented Indentation Test for Hardness and Materials Parameters—Part 1: Test Method*; International Organization for Standardization: Geneva, Switzerland, 2002.
31. Oliver, W.C.; Pharr, G.M. An improved technique for determining hardness and elastic modulus using load and displacement sensing indentation experiments. *J. Mater. Res.* **1992**, *7*, 1564–1583. [CrossRef]
32. Rho, J.Y.; Pharr, G.M. Nanoindentation testing of bone. In *Mechanical Testing of Bone and the Bone-Implant Interface*; An, Y.H., Draughn, R.A., Eds.; CRC Press: Boca Raton, FL, USA, 1999; pp. 257–269.
33. Iijima, M.; Muguruma, T.; Brantley, W.A.; Mizoguchi, I. Comparisons of nanoindentation, 3-point bending, and tension tests for orthodontic wires. *Am. J. Orthod. Dent. Orthop.* **2011**, *140*, 65–71. [CrossRef] [PubMed]
34. Lee, Y.K. Colour and translucency of tooth-colored orthodontic brackets. *Eur. J. Orthod.* **2008**, *30*, 205–210. [CrossRef] [PubMed]

35. Esan, T.A.; Olusile, A.O.; Akeredolu, P.A. Factors influencing tooth shade selection for completely edentulous patients. *J. Contemp. Dent. Pract.* **2006**, *7*, 80–87. [PubMed]
36. Seuss, S.; Lehmann, M.; Boccaccini, A.R. Alternating current electrophoretic deposition of antibacterial bioactive glass-chitosan composite coatings. *Int. J. Mol. Sci.* **2014**, *15*, 12231–12242. [CrossRef] [PubMed]
37. Kusy, R.P.; Tobin, E.J.; Whitley, J.Q.; Sioshansi, P. Frictional coefficients of ion-implanted alumina against ion-implanted beta-titanium in the low load, low velocity, single pass regime. *Dent. Mater.* **1992**, *8*, 167–172. [CrossRef]
38. Burrow, S.J. Friction and resistance to sliding in orthodontics: A critical review. *Am. J. Orthod. Dent. Orthop.* **2009**, *135*, 442–447. [CrossRef] [PubMed]
39. Hu, W.; Featherstone, J.D. Prevention of enamel demineralization: An in-vitro study using light-cured filled sealant. *Am. J. Orthod. Dent. Orthop.* **2005**, *128*, 592–600. [CrossRef] [PubMed]
40. Gorton, J.; Featherstone, J.D. In vivo inhibition of demineralization around orthodontic brackets. *Am. J. Orthod. Dentofacial Orthop.* **2003**, *123*, 10–14. [CrossRef] [PubMed]
41. De Almeida, P.D.V.; Grégio, A.M.; Machado, M.A.; de Lima, A.A.; Azevedo, L.R. Saliva composition and functions: A comprehensive review. *J. Contemp. Dent. Pract.* **2008**, *9*, 72–80.

Article

Aluminum Templates of Different Sizes with Micro-, Nano- and Micro/Nano-Structures for Cell Culture

Ming-Liang Yen [1], Hao-Ming Hsiao [2], Chiung-Fang Huang [3,4], Yi Lin [5], Yung-Kang Shen [3,6,*], Yu-Liang Tsai [6,7], Chun-Wei Chang [8], Hsiu-Ju Yen [9], Yi-Jung Lu [9] and Yun-Wen Kuo [9]

1 Division of Oral and Maxillofacial Surgery, Department of Dentistry, Taipei Medical University Hospital, Taipei 10617, Taiwan; abu1106@gmail.com
2 Department of Mechanical Engineering, National Taiwan University, Taipei 10617, Taiwan; hmhsiao@ntu.edu.tw
3 School of Dental Technology, College of Oral Medicine, Taipei Medical University, Taipei 110, Taiwan; d642078@yahoo.com.tw
4 Department of Dentistry, Taipei Medical University Hospital, Taipei 110, Taiwan
5 Department of Business Administration, Takming University of Science and Technology, Taipei 114, Taiwan; linyi@takming.edu.tw
6 Research Center for Biomedical Devices, Taipei Medical University, Taipei 110, Taiwan; b10204102@mail.ntust.edu.tw
7 Department of Materials Science and Engineering, National Taiwan University of Science and Technology, Taipei 106, Taiwan
8 Division of Endodontic, Department of Dentistry, Taipei Medical University Hospital, Taipei 10617, Taiwan; ash62612@yahoo.com.tw
9 Division of Family and Operative Dentistry, Department of Dentistry, Taipei Medical University Hospital, Taipei 10617, Taiwan; b202093069@tmu.edu.tw (H.-J.Y.); yi_jung2002@yahoo.com.tw (Y.-J.L.); kjpopo@hotmail.com (Y.-W.K.)
* Correspondence: ykshen@tmu.edu.tw; Tel.: +886-2-2736-1661 (ext. 5147); Fax: +886-2-2736-2295

Academic Editor: Saber AminYavari
Received: 23 August 2017; Accepted: 20 October 2017; Published: 26 October 2017

Abstract: This study investigates the results of cell cultures on aluminum (Al) templates with flat-structures, micro-structures, nano-structures and micro/nano-structures. An Al template with flat-structure was obtained by electrolytic polishing; an Al template with micro-structure was obtained by micro-powder blasting; an Al template with nano-structure was obtained by aluminum anodization; and an Al template with micro/nano-structure was obtained by micro-powder blasting and then anodization. Osteoblast-like cells were cultured on aluminum templates with various structures. The microculture tetrazolium test assay was utilized to assess the adhesion, elongation, and proliferation behaviors of cultured osteoblast-like cells on aluminum templates with flat-structures, micro-structures, nano-structures, and micro/nano-structures. The results showed that the surface characterization of micro/nano-structure of aluminum templates had superhydrophilic property, and these also revealed that an aluminum template with micro/nano-structure could provide the most suitable growth situation for cell culture.

Keywords: surface modification; micro-powder blasting; aluminum anodization; micro/nano-structure; cell culture

1. Introduction

The surface of dental- or bone-implanted objects must commonly be modified to yield a particular surface roughness in order to increase their surface area for osteoblast attachment, and to enhance the bioactive and osteoconductive properties of the underlying substrate. Effective surface treatment methods include sand- or grit-blasting using abrasives, chemical treatments, and the deposition of calcium phosphate (CaP) coatings.

Technological developments have enabled the preparation of nano-scale structures, including anodized aluminum with neat arrays of holes known as porous alumina, which is a biomedical material. Biomedical engineering involves cell culture, biomedical materials, and surface modification. The cell growth is improved by a biomaterial with an effective structure. Numerous scholars are interested in the scale, micro-structure, and nano-structure of biomaterials.

The powder blasting method for hydroxyapatite (HA) was utilized to blast on a pure titanium (Ti) substrate. They found that the content and crystal structure of Ti substrate after blasting were the same as those of pure HA. The bonding strength of Ti substrate after powder blasting was larger than that by the dip coating, electrolysis deposition, and electrochemical deposition [1]. An animal test was performed for pure Ti after surface modification by blasting. The results demonstrated that the thickness of new bone on Ti substrate after being HA blasted exceeded that of pure Ti substrate [2]. A new method was developed for blasting a Ti surface that involved aluminum oxide (Al_2O_3) and a dopant (HA, fluoro apatite (FA), magnesium apatite (MgA) and carbonate apatite (CO_3A)), and a cell culture was performed on that surface. The results indicated the greatest proliferation of cells on the Ti substrate that was blasted by Al_2O_3 and CO_3A particles [3]. The biocompatibility of Ti substrate was discussed on the condition of being treated by HA blasting alone and by HA blasting with Al_2O_3. The results revealed that the surface roughness of the Ti substrate was greater following Al_2O_3 treatment and HA blasting. A cell culture revealed that the viability of cells on Ti substrate that was treated with Al_2O_3 followed by HA blasting exceeded that of the substrate that had undergone only HA blasting. The results also revealed that the growth of laminate bone has good situation on the Ti substrate that was treated with Al_2O_3 and HA blasting [4]. The antibacterial effectiveness of Ti substrate treated with pure HA particles or HA combined with zinc apatite (ZnA), silver apatite (AgA), or strontium apatite (SrA) particles were evaluated, and it was found that the substrate that was treated with HA and AgA performed best in this respect [5]. The wear and friction of a $TiAl_6V_4$ substrate that was combined with Al_2O_3 and teflon, silicon carbide (SiC), or boron carbide (B_4C) by blasting method was investigated [6]. An MG63 cell culture was conducted on a Ti substrate after blasting with HA and sintered CaP particles. The results revealed that surface modification increased cell proliferation on the Ti substrate [7]. A MG63 cell culture was carried out in vitro on the $TiAl_6V_4$ following surface modification (using different co-blasting methods). Their results demonstrated that co-blasting with bioglass and HA particles improved the osteoconduction and growth of cells on $TiAl_6V_4$. Their results also indicated that co-blasting of the $TiAl_6V_4$ substrate yielded a better alkaline phosphatase (ALP) value than the plasma spray method [8,9]. The researchers reviewed 348 papers on MG63 proliferation on Ti and $TiAl_6V_4$ substrates that had undergone various methods of surface modification [10]. The nanostructure of substrate affected the adhesion and proliferation of cells in vitro. The results showed that the moderately rough substrates with large fractal dimension could boost cell proliferation [11,12]. The morphology and biocompatibility of polished nitinol (NiTi) and Ti material surfaces treated with a mixed solution of three acids (HCl–HF–H_3PO_4) were evaluated. The results showed that surfaces treated with HCl–HF–H_3PO_4 had higher roughness, lower cytotoxicity, and better biocompatibility than controls [13,14]. MG63 cells were seeded on machined pretreated, nano-modified pretreated, sandblasted/acid-etched, and nano-modified sandblasted/acid-etched Ti disks. The results revealed that the nanoscale structures in combination with micro-/submicro-scale roughness improved osteoblast differentiation and local factor production, which indicated the potential for improved implant osseointegration [15,16].

The oxidation of anodic aluminum oxide (AAO) in sulfuric acid, phosphoric acid, or oxalic acid yields anodized porous alumina. Generally, AAO has a highly porous array structure and straight uniform pores. The diameter of pores varies with anodic reaction conditions. Straight nano-channels of AAO are often used to provide a framework for the formation of highly regular nano-structured materials. Porous anodic alumina membranes are formed from metal aluminum in acidic solution by two-step anodization [17–23]. The most commonly used electrolytes are sulfuric acid, oxalic acid, and phosphoric acid solution. In the anodizing process, aluminum is the anode, an electric field is applied, and the surface of the aluminum forms an oxide film. The extent of electrolytic oxidation increases with the voltage. Varying the electrolyte and the electrolysis time yields alumina membranes with pore diameters up to several hundred nanometers, or as small as 5 nm. The hole density up to 1011 holes/cm [24–27] and film thickness from 10 to 100 μm can be obtained. The porous alumina template is by far the most widely used template because it has monodispersed characteristic, it can resist high temperatures, and has high strength. The resulting nanotopography combines ordered nanostructures with widely varying surface energies, providing a unique platform for studying cell–substrate interactions. Human dermal fibroblasts were cultured on these substrates. Surface patterning with nanoscale pillars markedly affected cell morphology, which was independent of surface energy. Cell spreading was significantly reduced on both hydrophobic and hydrophilic surfaces with nanopillars. This analytical result shows that surfaces which resist cell spreading can be fabricated by generating suitable nanoscale topography, without concern for the effect of surface chemistry on hydrophilicity [28–31]. Popat et al. [32] established that the cell activity on AAO exceeded that on pure aluminum. Hoess et al. [33] showed that the filopodia of a HepG2 cell passed through nanoholes with a diameter of 263 nm, favoring cell adhesion and proliferation.

The motivation of this study is to study the cell culture on aluminum templates with various structures for application on dental- or bone-implanted objects. The purpose of this study is to discuss the behaviors of cell culture on the various structures of Al template by different surface modification methods. The authors have developed the mini screw on prosthodontics in Taipei Medical University. The mini screw was used as the aluminum material. The research emphasizes that the surface property of the mini screw (as the implanted object) influences the osseointegration. This investigation concerns cell culture on aluminum templates with flat structures, micro-structures, nano-structures, and micro/nano-structures. This study focuses on the various structures of Al templates for osteoblast-like cells (MG63, human osteosarcoma cell), because these cells (MG63) have different effects on aluminum templates of micro-sized structures formed by micro-powder blasting and on aluminum templates of anodized nanometer-sized structures and on aluminum templates of micro/nano-structures by micro-powder blasting and anodized process. This study emphasizes the surface roughness and surface property (hydrophilic or hydrophobic) on aluminum templates with various structures for cell culture. The purpose of this study is to apply the implanted object for bone or teeth to osseointegration. This can improve the stability of bone or dental implanted objects and decrease the repair time of bone or teeth. The null hypothesis is that the surface modification methods (micro-powder blasting, anodized process, micro-powder blasting + anodized process) only have an effect on the surface properties of the Al template.

2. Materials and Methods

2.1. Materials

Specimens were prepared from circular aluminum (Al) templates (99.9%, thickness = 1 mm, Φ = 15 mm) using various processes. To prepare a flat-structure specimen, the Al template was electropolished in a solution of perchloric acid and ethanol ($HClO_4$:C_2H_5OH = 1:4) at 7 °C for 2 min to remove surface irregularities.

2.2. Micro-Structure of Al Template

A micro-structured Al template was prepared. A micro-blasting machine (MICROPEEN 1300 ZP/ZPD, Iepco, Leuggern, Switzerland) was used to perform micro-powder blasting on Al template (99.9%). The micro-powder blasting formed irregular concave micron-sized holes on the surface of the aluminum template. This method increased the surface roughness of the aluminum. The sands used for micro-powder blasting were MS 245A, MS 300A, MS 550A and MS 550BT (A means sharp sand,

BT means round sand), with a diameters of 50–250, 30–70, 10–20, and 20–30 microns, respectively. To form an aluminum template with micro-structure, a blasting pressure of six bars was utilized; the distance between the blasting nozzle and the template was 3 cm; two blasting times were used in each case, and four kinds of sand particles were used.

2.3. Nano-Structure of Al Template (AAO)

A nano-structured Al template was formed, and anodic aluminum oxide (AAO) was prepared as follows:

- Pre-treatment: Aluminum template with a purity of 99.9% was soaked in an alcohol solution and ultrasonically vibrated for 30 min. It was then placed in 5% NaOH and soaked for 3 min to remove surface oil. Following heat treatment (400 °C, 4 h), it was electrolytically polished using 85% perchloric acid ($HClO_4$, Kanto Chemical Co., Ltd., Tokyo, Japan) and 15% ethanol (C_2H_5OH). It was then washed twice in deionized water.
- Anode handling: (1) Aluminum template was firstly anodized using 0.5 M oxalic acid on 30 V at room-temperature for 1 h to do the anodic process for the first time; (2) Chemical etching: The aluminum was rinsed for the second time in deionized water, and placed in a solution of 1.5 wt % chromic acid (Katayama reagent Co., Ltd., Osaka, Japan) that had been mixed with 6 wt % phosphoric acid (Katayama reagent Co., Ltd.) at 70 °C. The reaction time was 1 h. The growth was etched to retain a few pit holes under its surface. It was then washed twice in deionized water, before being anodized for the second time; (3) The second anodic treatment was conducted using 0.5 M oxalic acid at 30 V and room temperature for 3 h.

2.4. Micro/Nano-Structure of Al Template

To generate a surface with micro/nano-structure, the micro-powder blasting method and an anodization process (voltage: 30 V, 0.5 M oxalic acid, room temperature, first anodized period time: 1 h, second anodized period time: 3 h) were utilized to form nano-holes in a micro-structured aluminum template. The novelty of this work is the use of an innovative method to fabricate an Al template with micro/nano-structure. This method is easy, fast, and cheap for the production of the micro/nano-structured Al template. The surface properties of the flat aluminum or aluminum templates with various structures (micro-structure, nano-structure, and micro/nano-structure) importantly affect the cell culture that is performed on these materials. Figure 1 displays the fabrication of Al template with various structures.

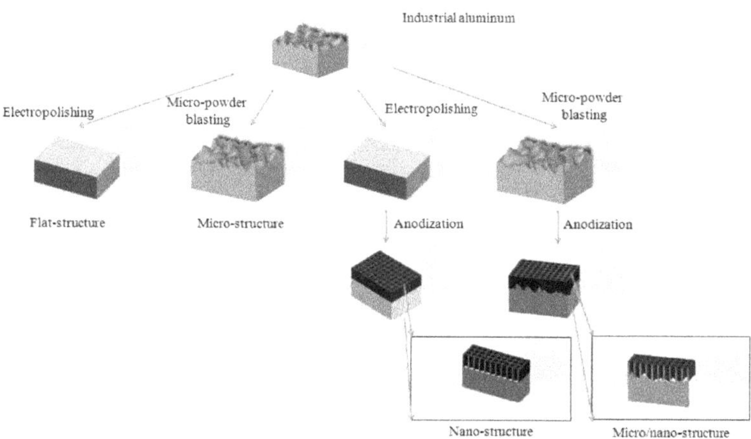

Figure 1. Fabrication process for different structures of aluminum (Al) templates.

2.5. Surface Properties

A contact angle meter (DIGIDROP DGD-DI, GBX, Dublin, Ireland) was used to measure the contact angle of aluminum templates with various structures. The contact angles of the surfaces of Al template with flat-structure, micro-structure, nano-structure, and micro/nano-structure are discussed. 5 points were measured on each specimen. Deionized water (0.5 μL) was dropped on the template surface. The three states of solid/gas/liquid affected the liquiddrop stability, the use of computer-controlled photography (25/s) captured images and converted the image files. The obtained data of measured contact angle were made into charts.

A MultiMode 3D scanning probe atomic force microscope (DI 3100, Advanced Surface Microscopy, Indianapolis, IN, USA) was utilized to determine the surface roughness of Al templates with various structures. The atomic force microscope (AFM) was also applied to measure the surface profile of aluminum templates with different structures. The authors measured the surface roughness of each test template at 5 measurement points. The scanning range of each measurement point was 5×5 μm^2. The surface morphology of Al templates with various structures was analyzed by SEM (JSM-6700F, JOEL, Peabody, MA, USA).

2.6. Cell Culture

In the phosphate-buffered saline (PBS) buffer allocation method, PBS and deionized water were mixed in ratio of 1:9. The PBS was put in a sterilized bottle and stored in a refrigerator at 4 °C. Dulbecco's Modified Eagle Medium from HyClone Co. (South Logan, UT, USA) was added to 10% PBS and 1% penicillin (HyClone Co.). The MG63 cell line (ATCC CRL-1427) was used in the cell culture. MG63 is a human bone precursor cell (human osteogenic sarcoma). MG63 cells are utilized in experimental research of the in vitro attachment and proliferation of bone cells. The microculture tetrazolium test (MTT) is 3-(4,5-cimethylthiazol-2-yl)-2,5-diphenyl tetrazolium bromide. It is a yellow compound that accepts hydrogen ions. It acts on the respiratory chain of living cell lines. Cracking its tetrazolium ring using succinate dehydrogenase and cytochrome C yields a blue formazan crystal. In this study, the crystal was dissolved in dimethyl sulfoxide, and its optical density (OD) was measured using an ELISA machine (Anthos 2020, Biochrom, Cambridge, UK). The OD value indicated the cell activity.

2.7. Statistics

Measured data were subjected to statistical analysis. For any given experiment, each data point represented the mean ± standard deviation (SD) of six individual experiments. The Tukey-test was used to determine significance between groups in the contact angle and surface roughness. Statistical significance was indicated by * $p < 0.05$, ** $p < 0.01$, and *** $p < 0.001$.

3. Results and Discussion

3.1. Surface Morphology of Micro-Structure of Al Template

In this work, micro-powder blasting was carried out to form a micron-scale surface on an aluminum template, which was then anodized to produce nanoholes in anodic alumina. Finally, the micro/nano-structure of the aluminum template was formed. The first goal was to form a suitable micro-structured surface of aluminum using various sand particles on micro-powder blasting (Figure 2). The depth of the surface of the aluminum micro-structure declined as the diameter of the sand particles declined (Figure 2a–c). The results also reveal that surfaces which had been impacted by larger sand particles were more concave and convex. Figure 2d demonstrates that the depth of the surface micro-structure of aluminum that underwent impact by round sand particles was less than that which underwent impact by sharp sand particles. The results also indicate that the micro-structure surface of aluminum that underwent impact by round sand particles was smoother than that by sharp sand particles.

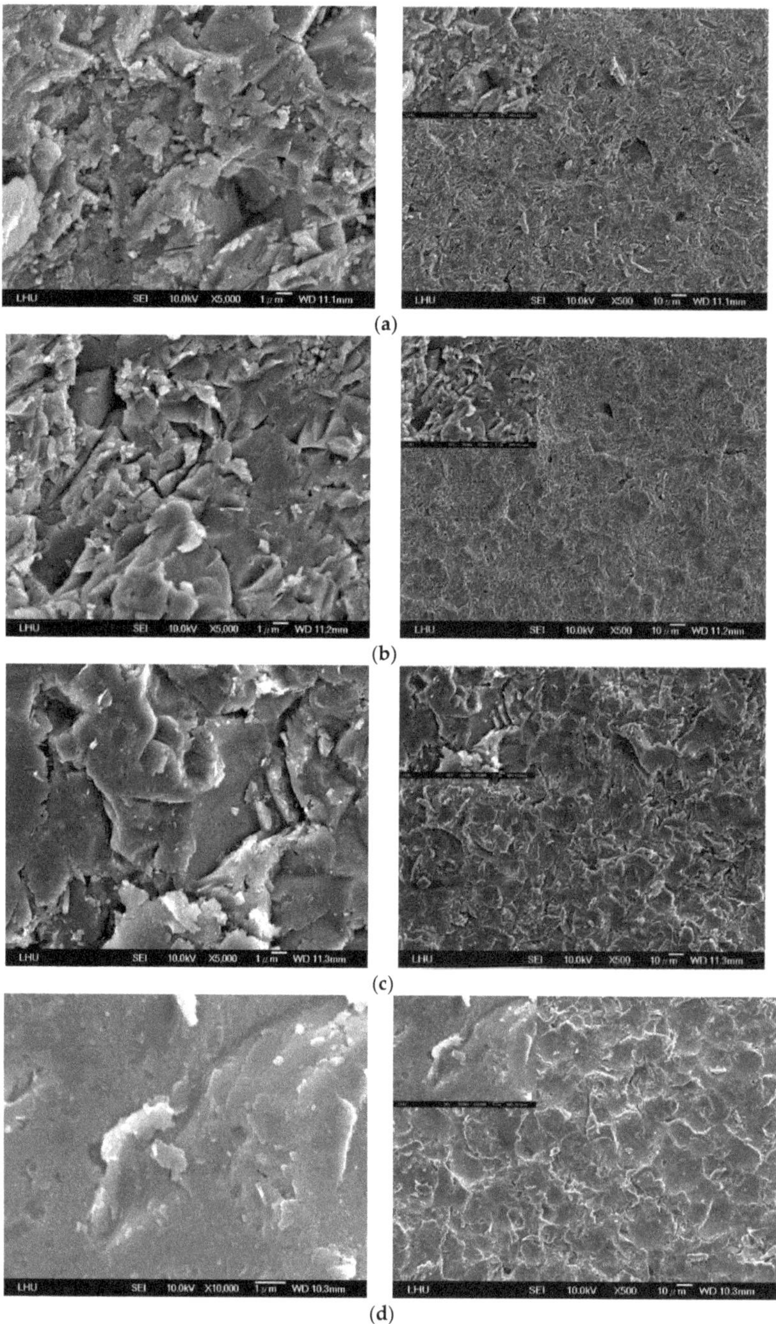

Figure 2. SEM images of Al template by powder blasting processing: (**a**) MS 245A (Φ = 50–250 μm); (**b**) MS 300A (Φ = 30–70 μm); (**c**) MS 550A (Φ = 10–20 μm); (**d**) MS 550BT (Φ = 20–30 μm).

3.2. Surface Morphologies of Nano and Micro/Nano-Structures of Al Template

To assess the quality of the prepared anodized aluminum templates, they were observed using SEM and AFM, as presented in Figure 3. Figure 3a shows the SEM images of AAO that were formed by the anodization process. The results reveal that the mean pore size in AAO was approximately 100 nm. Additionally, the formed pore arrays of AAO were very uniform. The micro/nano-structure of the Al template obtained by micro-powder blasting and anodic oxidation process is discussed. Figure 3b presents an SEM image for the previous process with electrolytic polishing, followed by the anodic oxidation process. The results demonstrate that micro-powder blasting barely formed a micro-structure, but rather formed nanoholes in AAO, yielding a pore size of about 60–80 nm. Figure 3c shows the SEM images following anodic oxidation process without electrolytic polishing. The results indicate that the micro-structure formed on the Al template, and nanoholes formed in the micro-structure with sizes about 50–80 nm, suggesting that the micro/nano-structure formed on the Al template.

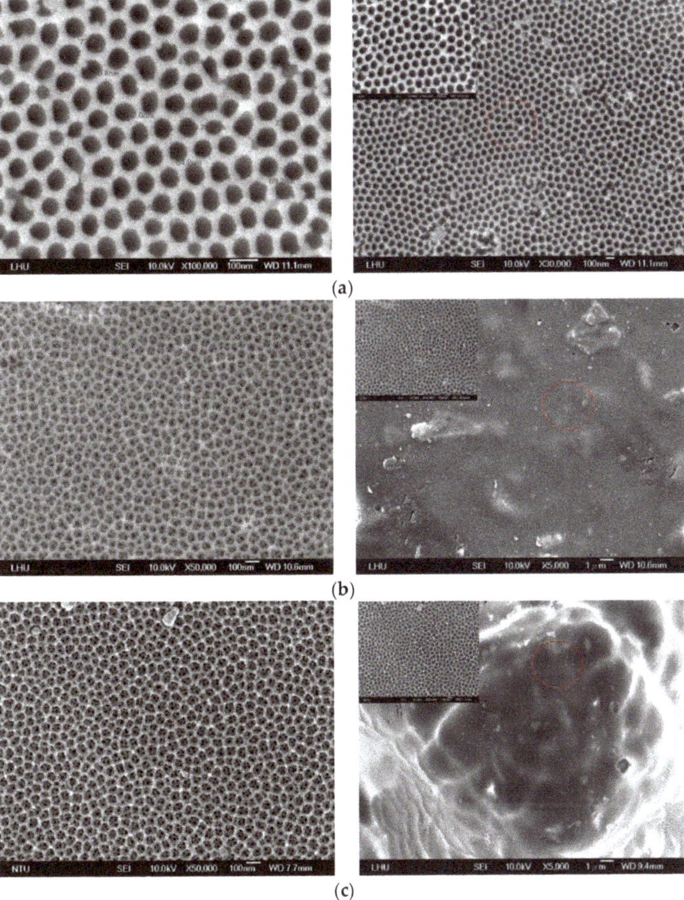

Figure 3. SEM images of nano-structure and micro/nano-structure: (**a**) Φ = 100 nm (anodic aluminum oxide (AAO)); (**b**) Φ = 60–80 nm (nano-hole on micro/nano-structure with electrolysis polishing); (**c**) Φ = 50–80 nm (nano-hole on micro/nano-structure, without electrolysis polishing).

3.3. Surface Properties of Various Al Template Structures

The surface properties of the various structured templates importantly affect the cell culture thereon. The effects of contact angle and surface roughness of Al templates with various structures on their surface are considered. Table 1 displays measured contact angles on smooth Al template (flat-structure), a micro-structured Al template that was formed by micro-powder blasting (MS 245A, MS 300A and MS 550BT), the nano-structure (AAO) and the micro/nano-structured Al template, revealing that the contact angles of the Al templates with the different structures fall into three groups. The contact angles of the smooth Al template and the micro-structure of Al substrate formed by micro-powder blasting (MS 245A) were about 77°–88°. The contact angle of Al template by MS 245A did not decline very much as the size of the sand particles increased, yielding a larger micro-structure, so the surface properties of the Al template did not improve with an increase in the particle size. The contact angles of the Al template with micro-structure formed by micro-powder blasting (MS 300A, MS 550BT) were around 28°–36°, revealing that the surface of the Al template by micro-powder blasting changed from hydrophilic to more hydrophilic. Furthermore, the contact angles of Al templates with various structures were affected by the size of sand particles, and are independent of their shapes. Finally, the contact angles of the nano-structure (AAO) and micro/nano-structure on Al template were about 7°–21°, indicating that these structures are superhydrophilic. These results also show that the micro/nano-structured Al template had the lowest contact angles, and that the micro/nano-structure of the Al template was more hydrophilic than the other structures of the Al template. Contact angle values by Tukey-test are also listed in Table 1. The contact angle indicates that there was no statistically significant difference between the smooth Al template and the Al template with micro-structure (MS 245A). The results also show that the contact angle between Al template with micro-structure (MS 300A) and the Al template with micro-structure (MS 500BT) had no statistically significant difference. The other two had statistically significant differences from each other in terms of contact angle for different structured templates.

Table 1 also presents the surface average roughness (R_a) values of Al templates with various structures. The results of surface roughness (R_a) indicate that the R_a of the Al template after micro-powder blasting is larger than that of flat Al template. The Wenzel equation appears that the surface roughness of the template can improve the surface wetting property. The contact angle of the template decreased as its surface roughness increased [34]. The results indicate that micro-powder blasting with smaller sand particle yielded larger R_a values of the Al template, because larger sand eroded the Al template more strongly and it could not produce small bumps on the surface of the Al template. The R_a value of the Al template fell as the size of the sand in the micro-powder blasting increased. The results also demonstrated that rounder sand in micro-powder blasting yielded smaller R_a values of the Al template, because sharp sand could more easy produce bumps on the surface of the Al template. The micro/nano-structure of the Al template had the highest R_a value. The Tukey-test for surface roughness of Al templates with various structures is also listed on Table 1, indicating that there was no statistically significant difference between the nano-structured (AAO) template and the micro-structured template (MS 245A). The results also reveal that the surface roughness between the micro-structured template (MS 300A) and the micro/nano-structured template had no statistically significant difference. The other two had statistically significant differences with each other in surface roughness for different structured templates.

The results of this study also reveal that the micro-powder blasting + anodized method yielded the minimum contact angle and the maximum surface roughness in the Al template. The contact angle had a smaller value and the surface roughness had the smallest value via the anodized method. The contact angle was largest via micro-powder blasting with different sized particles.

Table 1. Contact angles and surface roughnesses for different structured templates by Tukey-test.

Group	Contact Angle (°)	Surface Roughness (nm)	p-Value	Tukey-Test
AAO	18.76 ± 3.09	12.65 ± 0.06	0.001 ***/0.001 ***	A/A
MS 245A	80.70 ± 3.86	13.88 ± 0.07	–	B/A
MS 300A	33.58 ± 3.04	51.67 ± 0.13	–	C/B
MS 550BT	31.22 ± 3.02	40.07 ± 0.20	–	C/C
Micro/nano-structure on Al substrate	11.08 ± 4.19	56.37 ± 0.28	–	D/B
Smooth Al	85.72 ± 3.54	25.57 ± 0.13	–	B/D

*** $p < 0.001$.

3.4. Cell Viability Evaluation in Vitro

Figure 4 displays the MTT assay on Al templates with various structures. The OD value was statistically significantly different between the smooth Al template and the micro/nano-structured Al template. The results show that the OD value increased with the cell time regardless of the template structure. The results also reveal that the micro/nano-structure of the Al template had the highest OD value because it had the largest surface area; this explains why its surface approaches superhydrophilicity. The results also show that the OD values among the smooth Al template and micro/nano-structured Al template were statistically significantly different at day 4. In vitro studies revealed that the growth response of specific cell types give insight into the surface properties of the substrate. The surface roughness affects the cell response. The growth behavior of osteoblast-like cells (MG63) demonstrates the phenotypic characteristics of roughness-dependence. The results herein demonstrate that surface roughness may play an important role in determining cell response [35–37]. The results also demonstrate that the OD value depends on the contact angle. There are many studies indicating that the suitable hydrophilic property of a template surface can improve the cell adhesion and spreading on the surface [35,38,39]. A smaller contact angle yields a larger OD value, indicating that the hydrophilic nature of the template favors the cell adhesion and proliferation.

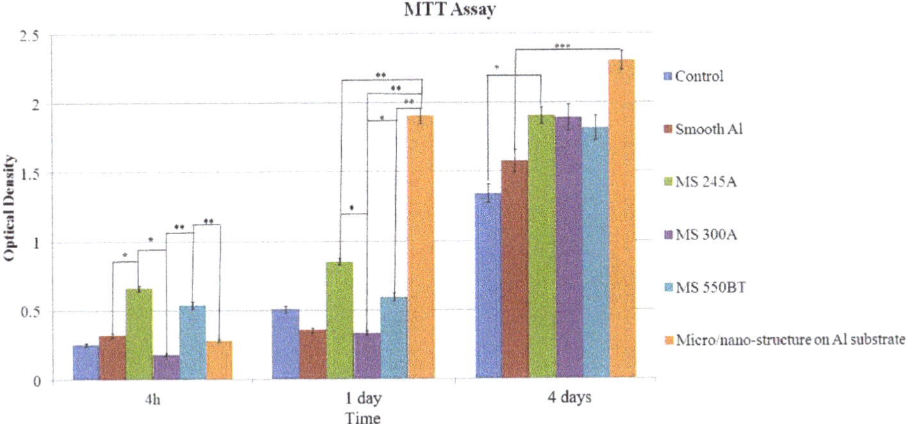

Figure 4. The microculture tetrazolium test (MTT) assay for different structured templates. (Values are the mean ± SD of six experiments ($n = 6$), * $p < 0.05$, ** $p < 0.01$, *** $p < 0.001$).

The results also indicate that a good behavior of cell adhesion and proliferation appeared on the surface of the Al template obtained by micro-powder blasting + anodized method, followed by the use of micro-powder blasting. The OD value had the smallest value on the smooth Al template.

3.5. The Results of Null Hypothesis

The null hypothesis was that surface modification methods (micro-powder blasting, anodized process, micro-powder blasting + anodized process) only has an effect on the surface of the Al template. The authors wanted to determine the depth of the Al template micro-structure or micro/nano-structure after surface modification. Figure 5 shows the surface profile of different structures of aluminum templates measured by AFM. The depth of the Al template was about 80.89 nm. The results show that the depths of the micro-structure of the Al template were 109.15 nm, 158.93 nm, and 164.52 nm for sand particles MS 245A, MS 300A, and MS 550BT by micro-powder blasting, respectively. The depth of AAO was 150.00 nm for the anodized process. The depth of Al template micro/nano-structure was 171.15 nm by the micro-powder blasting + anodized process. The previous results can reveal that the surface modification methods influence the surface layer of the Al template. The experimental results fit the null hypothesis.

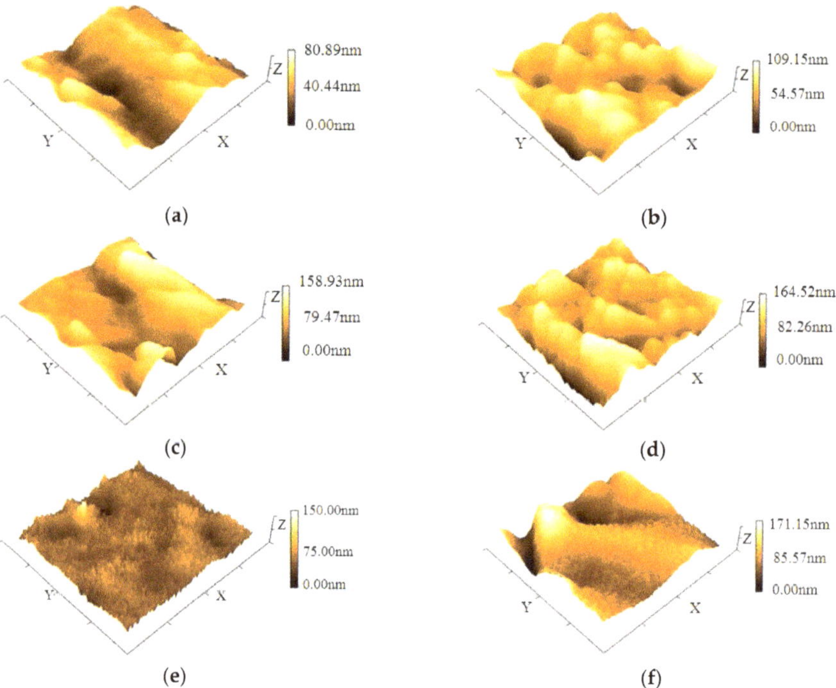

Figure 5. Surface profile of different structures on aluminum templates: (**a**) Al template; (**b**) MS 245A; (**c**) MS 300A; (**d**) MS 550BT; (**e**) AAO; (**f**) micro/nano-structure without electrolysis polishing.

4. Conclusions

This study evaluates the effects of Al template with various structures on cell cultures. The results can be used for the reference on bone or dental implants. The Al template with the flat-structure was slightly hydrophilic; the Al template with micro-structure formed by micro-powder blasting was more hydrophilic. The Al template with nano-structure became superhydrophilic by the anodization method. The Al template with micro/nano-structure became superhydrophilic, and it had the maximum value of contact angle. The Al template with micro/nano-structure had the maximum value of surface roughness, followed by the Al template with micro-structure, followed by the smooth Al template, and the Al template with nano-structure had the minimum value. Osteoblast-like cells (MG63) were cultured on the variously structured templates for 4 h, 1 day, and 4 days, before the MTT assays were performed. The results revealed that the Al template with micro/nano-structure had the highest OD value. The reason is that this template had the superhydrophilic property and the maximum surface roughness. The Al template with micro/nano-structure was more suitable for cell culture in this study.

Acknowledgments: The authors would like to gratefully acknowledge of financial support from National Taipei University of Technology-Taipei Medical University Joint Research Program (NTUT-TMU-101-12).

Author Contributions: Ming-Liang Yen and Yung-Kang Shen conceived and design the experiments; Hao-Ming Hsiao and Chiung-Fang Huang performed the experiments; Yi Lin, Yu-Liang Tsai and Chun-Wei Chang analyzed the data; Hsiu-Ju Yen, Yi-Jung Lu and Yun-Wen Kuo contributed analysis tools; Ming-Liang Yen and Yung-Kang Shen wrote the paper.

Conflicts of Interest: The authors declare no conflict of interest.

References

1. Ishikawa, K.; Miyamoto, Y.; Nagayama, M.; Asaoka, K. Blasting coating method: New method of coating titanium surface with hydroxyapatite at room temperature. *J. Biomed. Mater. Res. Part A Appl. Biomater.* **1997**, *38*, 129–134. [CrossRef]
2. Mano, T.; Ueyama, Y.; Ishikawa, K.; Suzuki, K. Initial tissue response to a titanium implant coated with apatite at room temperature using a blast coating method. *Biomaterials* **2002**, *23*, 1931–1936. [CrossRef]
3. O'Neill, L.; O'Sullivan, C.; O'Hare, P.; Sexton, L.; Keady, F.; O'Donoghue, I. Deposition of substituted apatites onto titanium surfaces using a novel blasting process. *Surf. Coat. Technol.* **2009**, *204*, 484–488. [CrossRef]
4. O'Hare, P.; Meenan, B.J.; Barke, G.A.; Byrne, G.; Dowling, D.; Hint, J.A. Biological responses to hydroxyapatite surface deposited via a co-incident microblasting technique. *Biomaterials* **2010**, *31*, 515–522. [CrossRef] [PubMed]
5. O'Sullivan, C.; O'Hare, P.; O'Leary, N.D.; Cream, A.M.; Ryan, K.; Dobson, A.D.W.; O'Neill, L. Deposition of substituted apatites with anticolonizing properties onto titanium surfaces using a novel blasting process. *J. Biomed. Mater. Res. Part B Appl. Biomater.* **2010**, *95*, 141–149. [CrossRef] [PubMed]
6. Fleming, D.; O'Neill, L.; Byrne, G.; Barry, N.; Dowling, D.P. Wear resistance enhancement of the titanium alloy Ti-6Al-4V via a novel coincident microblasting process. *Surf. Coat. Technol.* **2011**, *205*, 4941–4945. [CrossRef]
7. O'Sullivan, C.; O'Hare, P.; Byrne, G.; O'Neill, L.; Ryan, K.B.; Vrean, A.M. A modified surface on titanium deposited by a blasting process. *Coatings* **2011**, *1*, 53–71. [CrossRef]
8. Tan, F.; Naciri, M.; Dowling, D.; Rubeai, M.A. Osteoconductirity and growth factor production by MG63 osteoblastic cells on bioglass-coated orthopedic implants. *Biotechnol. Bioeng.* **2011**, *108*, 454–464. [CrossRef] [PubMed]
9. Tan, F.; Naciri, M.; Dowling, D.; Rubeai, M.A. In vitro and in vivo bioactivity of coblast hydroxyapatite coating and the effect of impaction on its osteoconductirity. *Biotechol. Adv.* **2012**, *30*, 352–362. [CrossRef] [PubMed]
10. Bachle, M.; Kohal, R.J. A systematic review of the influence of different titanium surface on proliferation, differentiation and protein synthesis of osteoblast-like MG63 Cells. *Clin. Oral Implant. Res.* **2004**, *15*, 683–692. [CrossRef] [PubMed]
11. Gentile, F.; Tirinato, L.; Battistam, E.; Causa, F.; Liberale, C.; Fabrizio, E.M.; Decuzzi, P. Cells preferentially grow on rough substrates. *Biomaterials* **2010**, *31*, 7205–7212. [CrossRef] [PubMed]
12. Decuzzi, P.; Ferrari, M. Modulating cellular adhesion through nanotopography. *Biomaterials* **2010**, *31*, 173–179. [CrossRef] [PubMed]
13. Zareidoost, A.; Yousefpour, M.; Ghaseme, B.; Amanzadeh, A. The relationship of surface roughness and cell response of chemical surface modification of titanium. *J. Mater. Sci. Mater. Med.* **2012**, *23*, 1479–1488. [CrossRef] [PubMed]
14. Yousepour, M.; Zareidoost, A. Evaluation of the effect of two-step acid etching on the surface treatment and improved bioactivity of nitinol. *J. Mash. Dent. Sch.* **2016**, *40*, 281–296.
15. Gittens, R.A.; Mclachlan, T.; Olivares-Navarrete, R.; Cai, Y.; Berner, S.; Tannenbaum, R.; Schwartz, Z.; Sandhage, K.H.; Boyan, B.D. The effects of combined micro-/submicro-scale surface roughness and nanoscale features on cell proliferation and differentiation. *Biomaterials* **2011**, *32*, 3395–3403. [CrossRef] [PubMed]

16. Gittens, R.A.; Olivares-Navarrete, R.; Schwartz, Z.; Boyan, B.D. Implant osseointegration and the role of microroughness and nanostructures: Lessons for spine implants. *Acta Biomater.* **2014**, *10*, 3363–3371. [CrossRef] [PubMed]
17. Wang, X.; Han, G.R. Fabrication and characterization of anodic aluminum oxide template. *Microelectron. Eng.* **2003**, *14*, 166–171. [CrossRef]
18. Yuan, J.H.; He, F.Y.; Sun, D.C.; Xia, X.H. A simple method for preparation of through-hole porous anodic alumina membrane. *Chem. Mater.* **2004**, *16*, 1841–1844. [CrossRef]
19. Firouzi, A.; Kumar, D.; Bull, L.M.; Besier, T.; Sieger, P.; Huo, Q.; Walker, S.A.; Zasadzinski, J.A.; Glinka, C.; Nicol, J. Cooperative organization of inorganic surfactants and biomimetic assemblies. *Science* **1995**, *267*, 1138–1143. [CrossRef] [PubMed]
20. Kresge, C.T.; Leonowicz, M.E.; Roth, W.J.; Vartuli, J.C.; Beck, J.S. Ordered malodorous molecular sieves synthesized by a liquid crystal templates mechanism. *Nature* **1992**, *359*, 710–712. [CrossRef]
21. Akmatsu, K.; Takeib, S.; Mizuhatab, M.; Kajinamib, A.; Dekib, S.; Takeokac, S.; Fujiid, M.; Hayashid, S.; Yamamotod, K. Preparation and characterization of polymer thin films containing silver and silver sulfide nano particles. *Thin Solid Film* **2000**, *359*, 55–60. [CrossRef]
22. O'Sullivan, J.P.; Wood, G.C. The morphology and mechanism of formation of porous anodic films on aluminum. *Proc. R. Soc. Lond. A* **1970**, *137*, 511–543. [CrossRef]
23. Keller, F.; Hunter, M.S.; Robinson, D.L. Structural features of oxide coatings on aluminum. *J. Electrochem. Soc.* **1953**, *100*, 411–419. [CrossRef]
24. Laet, J.D.; Anhellemont, J.; Terryn, H.; Vereechen, J. Characterization of various aluminum oxide layers by means of spectroscopic ellipsometry. *Appl. Phys. A.* **1992**, *54*, 72–78. [CrossRef]
25. Parkhutik, V.P.; Shershulsky, V.I. Theoretical modelling of porous oxide growth on aluminum. *J. Phys. D Appl. Phys.* **1992**, *25*, 1258–1263. [CrossRef]
26. Ng, K.O.; Vanderbilt, D. Stability of periodic domain structures in a two-dimensional dipolar model. *Phys. Rev.* **1995**, *2177*, 13–52. [CrossRef]
27. Chung, C.K.; Zhou, R.X.; Liu, T.Y.; Chang, W.T. Hybrid pulse anodization for the fabrication of porous anodic alumina films from commercial purity (99%) aluminum at room temperature. *Nanotechnology* **2009**, *20*, 005301–005304. [CrossRef] [PubMed]
28. Den, E.B.; Ruijter, E.D., Smits, J.H.; Ginsel, L.; Von, A.R.; Jansen, J. Quantitative analysis of cell proliferation and orientation on substrata with uniform parallel surface micro-grooves. *Biomaterials* **1996**, *17*, 1093–1099.
29. Clark, P.; Connolly, P.; Curtis, A.; Dow, J.; Wilkinson, C. Topographical control of cell behaviour: II. Multiple grooved substrata. *Development* **1990**, *108*, 635–644. [PubMed]
30. Wójciak, S.B.; Cutis, A.; Monaghan, W.; Macdonald, K.; Wilkinson, C. Guidance and activation of murine macrophages by nanometric scale topography. *Exp. Cell Res.* **1996**, *223*, 426–435. [CrossRef] [PubMed]
31. Clark, P.; Connolly, P.; Curtis, S.A.; Dow, A.J.; Wilkinson, C.D. Topographical control of cell behavior. I. simple step cues. *Development* **1987**, *99*, 439–448. [PubMed]
32. Popat, K.C.; Leary, E.E.S.; Mukhatyar, V.; Chatvanichkul, K.I.; Mor, G.K. Influence of nanoporous alumina membranes on long–term osteoblast response. *Biomaterials* **2005**, *26*, 4516–4522. [CrossRef] [PubMed]
33. Hoess, A.; Teuscher, N.; Thormann, A.H.; Heilmann, A. Cultivation of hepatoma cell line HepG2 on nanoporous aluminum oxide membranes. *Acta Biomater.* **2007**, *3*, 43–50. [CrossRef] [PubMed]
34. De Gennes, P.-G.; Brochard-Wyart, F.; Quere, D. *Capillarity and Wetting Phenomena*; Springer Inc.: New York, NY, USA, 2004.
35. Lampin, M.; Warocquier, R.; Legris, C.; Degrange, M.; Sigot-Luizard, M.F. Correlation between substrate roughness and wettabililty, cell adhesion and cell migration. *J. Biomed. Mater. Res.* **1997**, *36*, 99–108. [CrossRef]
36. Hallab, N.J.; Bundy, K.J.; O'Connor, K.; Moses, R.L.; Jacobs, J.J. Evaluation of metallic and polymeric biomaterial surface energy and surface roughness characterization for directed cell adhesion. *Tissue Eng.* **2001**, *7*, 55–71. [CrossRef] [PubMed]
37. Deligianmi, D.D.; Katsala, N.D.; Koutsoukos, P.G.; Missirlis, Y.F. Effect of surface roughness of hydroxyapatite on human bone marrow cell adhesion, proliferation, differentiation and detachment strength. *Biomaterials* **2001**, *22*, 85–96.

38. Van Wachem, P.B.; Bengelling, T.; Feijen, J.; Bantjes, A.; Detmers, J.P.; van Aken, W.G. Interaction of cultured human endothelial cells with polymeric surfaces of different wettabilities. *Biomaterials* **1985**, *6*, 403–408. [CrossRef]
39. Webb, K.; Hlady, V.; Tresco, P.A. Relative importance of surface wettability and charged functional groups on NIH 3T3 fibroblast attachment, spreading and cytoskeletal organization. *J. Biomed. Mater. Res.* **1998**, *41*, 422–430. [CrossRef]

Article

Effects of sp^2/sp^3 Ratio and Hydrogen Content on In Vitro Bending and Frictional Performance of DLC-Coated Orthodontic Stainless Steels

Takeshi Muguruma [1], Masahiro Iijima [1,*], Masahiro Kawaguchi [2] and Itaru Mizoguchi [1]

1 Division of Orthodontics and Dentofacial Orthopedics, School of Dentistry, Health Sciences University of Hokkaido, Hokkaido 061-0293, Japan; muguruma@hoku-iryo-u.ac.jp (T.M.); mizo@hoku-iryo-u.ac.jp (I.M.)
2 Surface Coating and Chemical Technology Group, Tokyo Metropolitan Industrial Technology Research Institute, 2-4-10 Aomi Koto-ku, Tokyo 135-0064, Japan; kawaguchi.masahiro@iri-tokyo.jp
* Correspondence: iijima@hoku-iryo-u.ac.jp; Tel.: +81-133-23-2977

Received: 2 April 2018; Accepted: 22 May 2018; Published: 24 May 2018

Abstract: This study investigated a diamond-like carbon (DLC) coating formed on stainless steels (disk and wire specimens) using a plasma-based ion implantation/deposition method with two different parameters (DLC-1, DLC-2). These specimens were characterized using high-resolution elastic recoil analysis, microscale X-ray photoelectron spectroscopy and nanoindentation testing to determine the hydrogen content, sp^2/sp^3 ratio and mechanical properties of the coating. Three-point bending and frictional properties were estimated. DLC-1 had a diamond-rich structure at the external surface and a graphite-rich structure at the inner surface, while DLC-2 had a graphite-rich structure at the external surface and a diamond-rich structure at the inner surface. Mean mechanical property values obtained for the external surface were lower than those for the inner surface in both types of DLC-coated specimens. The hydrogen content of DLC-2 was slightly higher versus DLC-1. Both DLC-coated wires produced a significantly higher elastic modulus according to the three-point bending test versus the non-coated wire. DLC-2 produced significantly lower frictional force than the non-coated specimen in the drawing-friction test. The coating of DLC-1 was partially ruptured by the three-point bending and drawing-friction tests. In conclusion, the bending and frictional performance of DLC-coated wire were influenced by the hydrogen content and sp^2/sp^3 ratio of the coating.

Keywords: diamond-like carbon; frictional property; hydrogen content; surface modification; sp^2/sp^3 ratio

1. Introduction

Metallic orthodontic appliances, such as brackets and archwires, typically show superior properties [1] and provide many clinical advantages, such as low frictional resistance and good bending performance as orthodontic archwires. They have been widely used in clinical orthodontics, although they have esthetic limitations compared to other orthodontic appliances made from ceramics and plastics. Another disadvantage of metallic orthodontic appliances is corrosion in the oral environment [2,3], because the release of metallic ions, such as nickel (Ni) and chromium (Cr), may cause an allergic reaction during orthodontic treatment [4–6].

The frictional force between the bracket and archwire (resistance to sliding) during tooth movement is a primary issue in orthodontics [7,8]. If the frictional force can be decreased, then the efficiency of the tooth movement can be improved. To improve the frictional characteristics and corrosion resistance, various surface modification techniques, such as diamond-like carbon (DLC) coating [9–12], plasma immersion ion implantation [7,13,14] and bioactive glass coating [15], have been investigated.

In recent years, DLC coating has become the subject of considerable research interest due to its bioinertness, extreme hardness, low friction coefficient and high wear resistance [16]. This technique has attracted significant attention for biomedical applications, such as artificial joints, cardiac stents and orthodontic archwires [17]. Concerning orthodontic applications, experimental DLC-coated orthodontic wires have been studied by several research groups [9–12,18–20]. One study reported that DLC layers protect against the diffusion of Ni and its release at the surface of Ni–Ti archwires and that these coatings are noncytotoxic in corrosive environments [18]. Other studies have investigated the effect of DLC coatings on the friction of orthodontic wires and found that DLC-coated wires produced less frictional resistance than non-coated wires [9–12,18–20]. The properties of a DLC coating depend on the hydrogen content, sp^2/sp^3 ratio and presence of doping elements [21,22]. The properties of DLC-coated orthodontic materials are not well understood, and limited information is available regarding the hydrogen content and sp^2/sp^3 ratio of DLC-deposited surfaces.

First, we deposited a DLC film onto orthodontic stainless steels using two different parameters and characterized the DLC films to determine their hydrogen content, sp^2/sp^3 ratio and mechanical properties. The bending and frictional properties of the DLC-coated orthodontic stainless steels were also investigated.

2. Materials and Methods

2.1. Materials

Mechanically-polished stainless steel disk specimens (diameter: 14 mm; thickness: 2 mm; Nogata Denki Kogyo, Tokyo, Japan) and as-received stainless steel orthodontic wires with cross-sectional dimensions of 0.017 × 0.025 in^2 (stainless steel archwire; 3M Unitek, Monrovia, CA, USA) were purchased and subjected to DLC coating. These stainless steels were confirmed to be Type 304 austenitic stainless steel (ISO No. 4301-304-00-I) by X-ray fluorescence analysis. As-received, preadjusted stainless steel orthodontic brackets (Mini Uni-Twin; 3M Unitek) for the upper canine teeth were used for friction tests. Non-coated specimens served as a control.

2.2. DLC Coating Procedure

DLC films were deposited onto stainless steel disks and wires using a plasma-based ion implantation/deposition (PBIID) method after the specimens were cleaned ultrasonically with acetone and alcohol. A custom-made jig was used to hold the specimens in the PBIID equipment (PEKURIS-HI; Kurita Seisakusho, Kyoto, Japan). To obtain DLC films with different compositions, two different parameters for target voltage, gas atmosphere and deposition time were used; these are listed in Table 1. All deposition processes were carried out at a pressure of 1.33 × 10^{-3} Pa.

Table 1. Deposition parameters for the DLC coating procedure used in the present study.

DLC Coating Procedure	Target Voltage	Gas Atmosphere	Deposition Time
DLC-1	10 kV	Acetylene + Toluene	3 min
DLC-2	7 kV	Toluene	4 min

2.3. Phase Identification by X-ray Diffraction and Scanning Electron Microscopy of the Coating

Wire specimens were cut into segments (length: 1 cm) using a water-cooled diamond saw (Isomet; Buehler, Lake Bluff, IL, USA). The segments were then placed side-by-side on the sample holder to yield ca. 1 × 1 cm^2 specimens. Representative surfaces of the control and DLC-coated wire specimens were analyzed using XRD (Rint-2500; Rigaku, Tokyo, Japan) via a parallel-beam method using Cu–Kα radiation (40 kV; tube current: 100 mA) over 2θ ranging from 10°–60° at a step size of 0.02° and a scan speed of 0.25° min^{-1}. The XRD patterns were obtained at 25 °C and analyzed for phase identification and quantification using PDXL2 software (Rigaku) based on the International Center for Diffraction Data (ICDD) database.

To observe the DLC-coated layers on a cross-sectioned surface, a wire specimen was encapsulated in an epoxy resin (Epofix; Struers, Copenhagen, Denmark) cross-sectioned with a slow-speed, water-cooled diamond saw (Isomet; Buehler) and then ground and polished using a series of silicon carbide abrasive papers and a final slurry of 0.05-μm alumina particles. All specimens were sputter-coated with pure gold for SEM evaluation (JSM-6610LA; JEOL, Tokyo, Japan); the SEM operated at 15 kV.

2.4. Compositional Characterization of the Coating by High-Resolution Elastic Recoil Analysis and Microscale X-ray Photoelectron Spectroscopy

An elastic recoil detection analyzer (ERDA; HRBS1000; Kobelco, Hyogo, Japan) was used for depth profiling of the hydrogen content of DLC-coated disk specimens. The ion type, acceleration voltage, incident angle and scattering angle were N^+, 500 kV, 67.5 and 45.6°, respectively. The main chamber was maintained at a pressure less than 1×10^{-5} Pa during the measurements. A multi-channel plate was used as the detector in this study. A beam of 500 keV N^+ ions was irradiated against the surface of the specimens, and hydrogen ions recoiled at 45.6° were measured by the 90° sector-type magnetic spectrometer. To reject the scattered N^+ ions, a Mylar foil was set in front of a multi-channel plate detector. The energy of hydrogen ions recoiled from the surface region of the implants was ca. 61 keV. Amorphous carbon materials with 20 at.% hydrogen were used as the standard sample. The standard sample was also measured under the same measurement conditions. The hydrogen contents of the specimens relative to carbon were compared with that of the standard sample. This enabled the depth profile of the contents to be calculated because the change in energy of the hydrogen ions corresponds to their depth from the surface.

The surface and in-depth composition of the control and DLC-coated disk specimens were analyzed by micro-XPS (Quantera II; Ulvac-Phi, Kanagawa, Japan) using Al Kα radiation with a 25-W beam power. The pressure of the main chamber was maintained at less than 1×10^{-6} Pa. Measurements on a 100 μm^2 area of the disk specimens were conducted from 0–1100 eV at a step size of 0.2 eV. The counting time was 20 ms for each step, and the number of sweeps was 5, i.e., the total counting time was 100 ms at each step. Argon-ion sputtering was used for depth profiling measurements. The ion sputtering area was 2×2 mm^2, and the measurements were taken at the center of the area. The sputtering rate of a SiO_2 layer under the same conditions was 13 nm min^{-1}. The sp^2 (for graphite) and sp^3 (for diamond) contents were determined using the software bundled with the XPS apparatus.

2.5. Mechanical Properties of the Coating from Nanoindentation and Three-Point Bending Testing

The external surfaces of DLC-coated wire specimens were examined with a nanoindentation apparatus (ENT-1100a; Elionix, Tokyo, Japan). The specimens were fixed to the specimen stage with adhesive resin (Superbond Orthomite; Sun Medical, Shiga, Japan). Nanoindentation testing was carried out at 28 °C using a Berkovich indenter for depth analyses at 20 and 70 nm (n = 10). Linear extrapolation methods (according to the ISO Standard 14577 [23]) were applied to the unloading curve between 95% and 70% of the maximum test force to calculate the elastic modulus. The hardness and elastic modulus of the wire specimen surfaces were calculated using the software bundled with the nanoindentation apparatus.

2.6. Evaluation of the Elastic Modulus of the DLC-Coated Wires by the Three-Point Bending Testing

A three-point bending test was carried out for non-coated and DLC-coated wires (n = 10). A 12-mm span was chosen for the wire segments in accordance with the ANSI/ADA Specification No. 32. All samples were loaded following the same protocol on a universal testing machine equipped with a 20 N load cell (EZ Test; Shimadzu, Kyoto, Japan) at room temperature (25 °C). Each wire was first loaded to a deflection of 1.0 or 1.5 mm and then unloaded at a rate of 0.5 mm min^{-1}. Following a three-point bending test, a specimen was inspected with a stereoscopic microscope (SMZ1500; Nikon, Tokyo, Japan) to observe the detachment of the DLC layers.

2.7. Frictional Properties Measured by the Progressive Load Scratch Test and Drawing Friction Test

A microtribometer (CETR-UMT-2; Bruker, Billerica, MA, USA) was used to characterize the frictional properties of each disk specimen by the progressive-load scratch test. A diamond stylus having a 12.5-μm tip radius was moved 5 mm over a specimen surface with linearly increasing normal load (0.5–20 gf) at a constant speed of 0.016 mm s^{-1}, and the value of the friction coefficient (tangential force) was obtained ($n = 5$). The initial frictional force, average frictional force during the first 0.5-mm scratch and total frictional force and average frictional force during the entire 5-mm scratch were calculated. After the scratch test, each specimen was inspected with a stereoscopic microscope (SMZ1500; Nikon, Tokyo, Japan) to determine the distance for detachment.

The forces generated with each wire/bracket combination were measured under dry and wet (in artificial saliva) conditions at room temperature (25 °C) using a custom-fabricated drawing-friction testing device attached to a universal testing machine (EZ Test; Shimadzu, Kyoto, Japan) [9]. Each bracket was bonded to a stainless steel plate with a non-filled adhesive resin (Superbond; Sun Medical, Shiga, Japan), and a bracket-mounting device provided 10° angular positioning for the bracket. The stainless steel plate with the bracket was attached to a friction-testing device. A 5-cm wire segment was then bound to the bracket using an elastic ligature (Alastik Easy-To-Tie Ligatures, 3M Unitek). The upper end of the wire was fixed to a grip attached to the load cell, and the lower end of the wire was fixed to a 150-g weight. Each wire was drawn through the bracket at a crosshead speed of 10 mm min^{-1} for a distance of 5 mm. The X axis was recorded for wire movement and the Y axis for the force. In the present study, the static frictional force was determined at the initial peak of movement, and the kinetic frictional force was calculated by averaging force values after the static friction peak [7,8]. The sample size for each condition was 10 ($n = 10$). After the drawing-friction test, a specimen was inspected with a stereoscopic microscope (SMZ1500; Nikon) to observe the detachment of the DLC layers.

2.8. Statistical Analyses

Statistical analyses were performed using SPSS Statistics software (ver. 23J for Windows; IBM, Armonk, NY, USA). The mean frictional forces, along with the standard deviation, were analyzed by two-way analysis of variance (ANOVA). The two factors were the coating procedure (non-coating, DLC-1, DLC-2) and test environment (dry, wet). Additionally, the mean hardness, elastic modulus and frictional force were compared using one-way ANOVA, followed by Tukey's or Games–Howell tests. The mean distance for detachment in the progressive-load scratch test was compared using Welch's t-test. For all statistical tests, significance was predetermined at $p < 0.05$.

3. Results

3.1. Crystal Structures and Morphological Features of the Coating Layers

Figure 1 displays representative XRD spectra of non-coated and DLC-coated wire specimens. No peak was obtained for either DLC-coated specimen due to their amorphous structures. The XRD spectra for the non-coated wire specimen contained peaks associated with the austenite phase (γ-Fe) (ICDD PDF 01-071-4649) and a non-indexed peak at 21.6°.

Representative SEM images of the non-coated and DLC-coated wire specimens are shown in Figure 2. The thin DLC layers on the wire specimen surfaces were ca. 300 nm thick for both the DLC-1 and DLC-2 cases. Good interfacial adhesion was observed between all DLC-deposited layers and bulk materials.

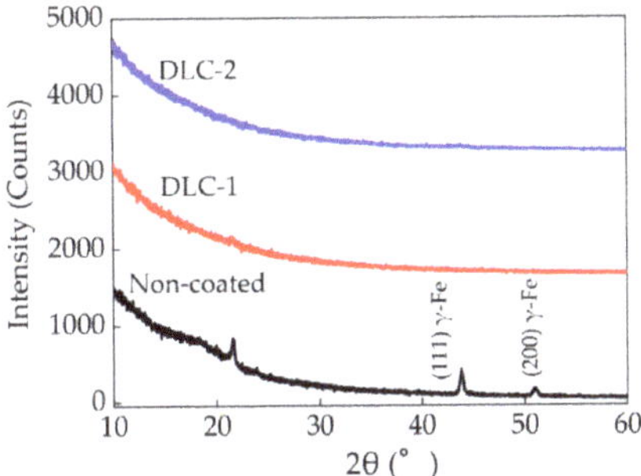

Figure 1. X-ray diffraction patterns obtained from the surfaces of non-coated and diamond-like carbon (DLC)-coated wire specimens.

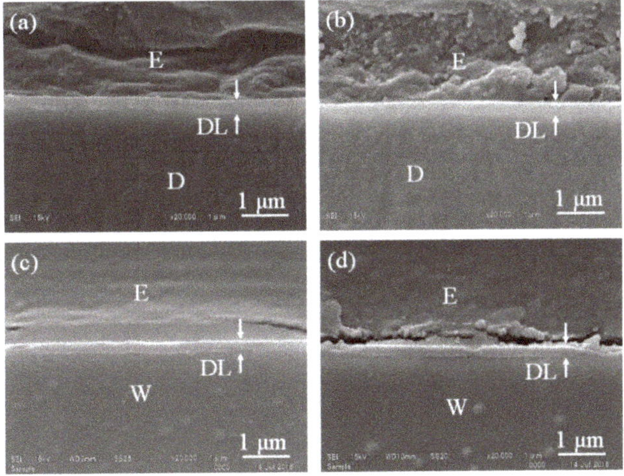

Figure 2. Scanning electron microscopy images of cross-sectioned disk (first raw) and wire specimens (second raw): (**a**,**c**) DLC-coated specimens (DLC-1) and (**b**,**d**) DLC-coated specimens (DLC-2). DL, DLC layer; D, disk; W, wire; E, epoxy resin. Original magnification: 20,000×.

3.2. Compositional Characterization of the Coating

Figure 3 shows hydrogen depth profile (concentration relative to carbon) by the elastic recoil detection analysis (ERDA) to a depth of 600 Å (60 nm) for the DLC-coated disk specimens. A higher hydrogen concentration was detected for DLC-2. The average hydrogen contents from the top surface to a depth of 600 Å were 23% for DLC-1 and 27% for DLC-2; the external surface regions contained 29% for DLC-1 and 33% for DLC-2.

Figure 4 shows the C 1s spectra obtained by XPS for the DLC-coated disk specimens. Gaussian-Lorentzian curve fitting was used to deconvolute the spectra into three peaks corresponding to sp^2 for graphite-like (284.5 eV) and sp^3 for diamond-like (285.3 eV) and CO-contaminated (283.56–288.43 eV). The amounts of sp^2 and sp^3 and the sp^2/sp^3 ratio (area) for each sputtered layer are summarized in Table 2 (a single layer was ca. 13 nm thick). The C 1s spectra almost disappeared from 40 layers because of the exposure of stainless steel surface to Ar-ion sputtering. The DLC-1 had a higher sp^2/sp^3 ratio (0.343) at the external surface region, although the value decreased (to 0.235) for four sputtered layers, which was similar to that for DLC-2 (0.283). On the other hand, DLC-2 had a graphite-rich external surface (sp^2/sp^3 ratio: 0.181), although the value increased (to 0.343) after nine sputtered layers, which indicated a diamond-rich surface.

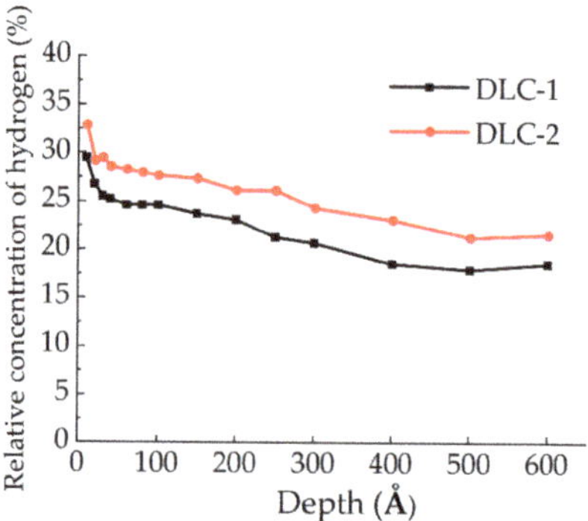

Figure 3. The hydrogen depth profile (concentration relative to carbon) to a depth of 600 Å (60 nm) determined from elastic recoil analyses of DLC-coated disk specimens.

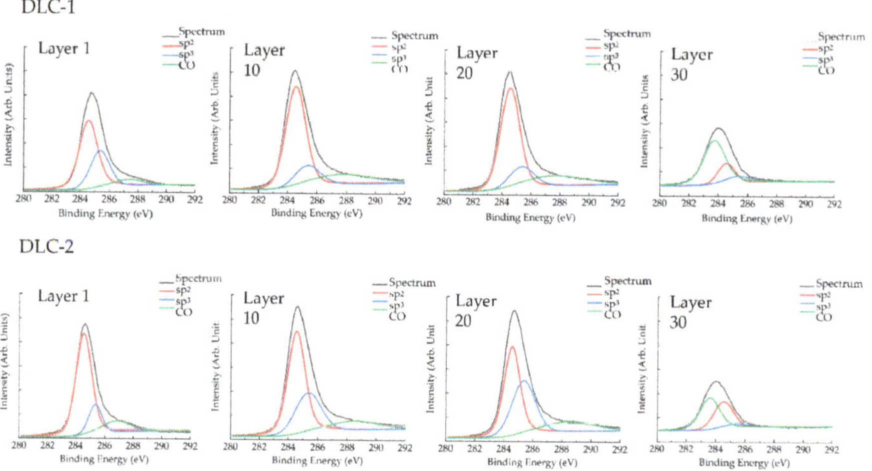

Figure 4. Gaussian-Lorentzian curve fitting of X-ray photoelectron spectroscopy C 1s spectra obtained for DLC-coated disk specimens.

Table 2. The sp^2, sp^3 and sp^2/sp^3 ratio for each sputtered layer.

Layers	DLC-1			DLC-2		
	sp^2	sp^3	$sp^3/(sp^2 + sp^3)$	sp^2	sp^3	$sp^3/(sp^2 + sp^3)$
1	51,573	26,885	0.343	64,025	14,128	0.181
5	82,776	25,388	0.235	80,737	31,883	0.283
10	78,206	19,654	0.201	73,440	38,368	0.343
15	74,470	20,126	0.213	63,305	45,916	0.420
20	76,467	20,132	0.208	60,451	47,915	0.442
25	73,992	20,656	0.218	71,608	23,626	0.248
30	12,557	6553	0.343	19,183	3356	0.149
35	3864	2553	0.398	911	2221	0.709
40	disappeared	disappeared	–	disappeared	disappeared	–

3.3. Mechanical Properties of the Coating

The mechanical properties of DLC-coated wire specimens obtained from nanoindentation testing at two analysis depths (ca. 20 and 70 nm) are summarized in Table 3. The mean values of the mechanical properties (hardness and elastic modulus) obtained for the external surface regions (at ca. a 20-nm depth) were lower than those for the inner surface regions (at ca. 70 nm depth) for both types of DLC-coated specimens. The DLC-1 tended to show higher mechanical properties at the external surface region and lower mechanical properties at the inner surface region compared with the DLC-2, although only the elastic modulus at the inner surface region was significantly different.

Table 3. Mechanical properties of DLC-coated wires obtained from nanoindentation testing (GPa).

Mechanical Properties	Analysis Depth	DLC-1	DLC-2	p Value [1]
Hardness	20 nm	8.13 (1.24)	7.49 (1.55)	0.317
	70 nm	9.18 (0.64)	9.69 (1.18)	0.241
Elastic modulus	20 nm	117.16 (19.59)	106.35 (36.60)	0.421
	70 nm	123.68 (6.42)	135.40 (12.04)	0.014

Notes: Values are presented as the mean ± SD; [1] Student t-test.

Table 4 summarizes the elastic modulus for the non-coated and DLC-coated wire specimens obtained by the three-point bending test. Both DLC-coated wires had a significantly higher elastic modulus (181–188 GPa) than the non-coated wire (170 GPa). For the 1.5-mm bending condition, the DLC-1 (188 GPa) had a significantly higher elastic modulus than the DLC-2 (181 GPa). Micrograph images taken following this three-point bending test revealed that the coating layer had been removed from the inner core for both DLC-coated wire specimens; none of the DLC-2 wire coatings were damaged after three-point bending at 1.0 mm (Figure 5).

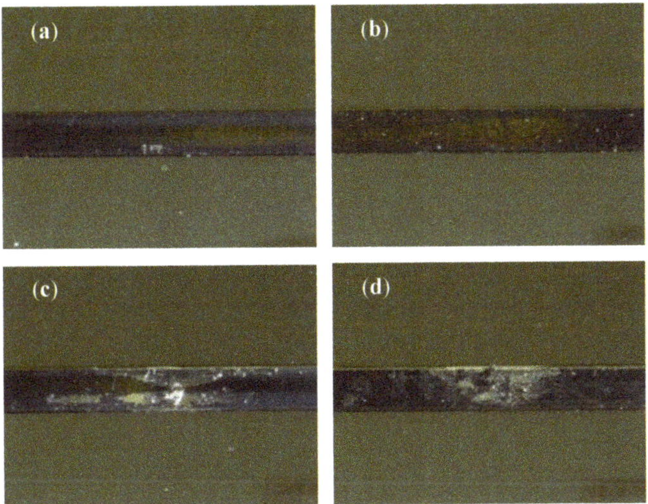

Figure 5. Stereomicroscope images of DLC-coated wires after the three-point bending test. (**a**,**c**) DLC-1 and (**b**,**d**) DLC-2. The first row shows specimens after the 1.0-mm bending and the second row specimens after the 1.5-mm bending. Original magnification: 50×.

Table 4. Elastic modulus for non-coated and DLC-coated wires obtained by the three-point bending test (GPa).

Bending	Non-Coated	DLC-1	DLC-2	p-Value
1 mm	170.45 ^{a} (1.87)	185.89 ^{b} (5.85)	181.84 ^{b} (3.50)	0.000
1.5 mm	170.26 ^{a} (2.18)	188.42 ^{b} (4.51)	180.80 ^{c} (2.54)	0.000

Notes: Values are presented as the mean ± SD; Identical letters indicate that mean values were not significantly different ($p < 0.05$) by one-way ANOVA followed by the Games–Howell test.

3.4. Frictional Properties Measured by the Progressive Load Scratch Test and Drawing Friction Test

Table 5 summarizes the frictional forces determined by the progressive-load scratch test. Both DLC-coated disk specimens had significantly lower initial and total frictional forces than the non-coated disk specimen. There was no significant difference between the two DLC-coated specimens in terms of the distance for detachment.

Table 5. Frictional forces obtained by the progressive-load scratch test (N).

Scratch Distances	Non-Coated	DLC-1	DLC-2	p-Value
5.0 mm	2.31 ^{a} (0.02)	2.01 ^{b} (0.03)	1.97 ^{b} (0.01)	0.000
0.5 mm	1.20 ^{a} (0.04)	1.03 ^{b} (0.07)	0.95 ^{c} (0.02)	0.000
Distance for detachment (mm)	–	1.05 (0.37)	1.27 (0.20)	0.269 †

Notes: Values are presented as the mean ± SD; Identical letters indicate that mean values were not significantly different ($p < 0.05$) by one-way ANOVA followed by the Tukey multiple test; † There was no significant difference between the two DLC-coated specimens in terms of the distance for detachment by Student's t-test.

Table 6 summarizes the static and kinetic frictional forces determined from drawing-friction testing of the non-coated and DLC-coated wire specimens under dry and wet conditions. Two-way ANOVA showed that the coating procedure (non coating, DLC-1, DLC-2) and test environment (dry, wet) were statistically-significant factors affecting both the static and kinetic frictional forces. One-way ANOVA and Tukey's tests showed that the DLC-2 had a significantly lower frictional force than the non-coated specimen, with the exception of the static frictional force under the dry condition. On the other hand, the DLC-1 showed frictional force values that were similar to those of the non-coated specimen, with the exception of the kinetic frictional force under the dry and wet condition. According to the drawing-friction testing, the DLC layers were partially ruptured for the DLC-1 case, while no rupture was observed for the DLC-2 condition (Figure 6).

Table 6. Static and kinetic frictional forces for the non-coated and DLC-coated wires in dry and wet conditions (N).

Friction Test	Condition	Non-Coated	DLC-1	DLC-2	p-Value
Static friction	Wet	2.39 ^{a} (0.30)	2.37 ^{a} (0.16)	2.09 ^{b} (0.22)	0.013
	Dry	2.49 (0.33)	2.47 (0.18)	2.25 (0.24)	0.088
Kinetic friction	Wet	2.37 ^{a} (0.27)	2.32 ^{a} (0.17)	1.99 ^{b} (0.17)	0.001
	Dry	2.55 ^{a} (0.21)	2.55 ^{a} (0.30)	2.21 ^{b} (0.18)	0.004

Notes: Values are presented as the mean ± SD; Identical letters indicate that mean values were not significantly different ($p < 0.05$) by one-way ANOVA followed by the Games–Howell test.

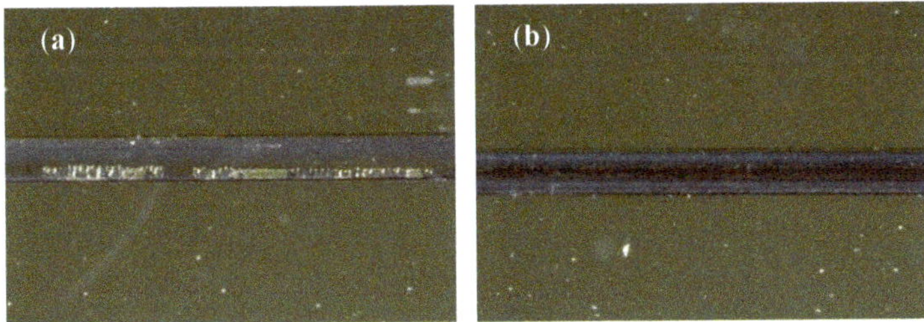

Figure 6. Stereomicroscope images of DLC-coated wires after the drawing-friction test. A portion of the DLC layer had ruptured from the interface. (**a**) DLC-1 and (**b**) DLC-2. Original magnification: 50×.

4. Discussion

In this study, ca. 300 nm-thick DLC layers were deposited on orthodontic stainless steels. The coatings were amorphous, which was consistent with previous findings [24]. The type of DLC can be identified using a ternary phase diagram [16]. This diagram shows the fraction of carbon sites that have sp^2 (graphite-like) bonding, sp^3 (diamond-like) bonding or bonding with hydrogen. Quantitative analysis of sp^2 and sp^3 bonding in a DLC can be performed by XPS analysis [25,26]. In the present study, the DLC-1 had a higher sp^2/sp^3 ratio (0.343) at the external surface region (ca. 13 nm deep), while the DLC-2 had a lower sp^2/sp^3 ratio (0.181) at the external surface region. This indicated that the external surface of the DLC-1 had a more diamond-rich structure than the DLC-2. After four more layers had been sputtered, the sp^2/sp^3 ratio (measured at a depth of ca. 65 nm) was similar for DLC-1 (0.235) and DLC-2 (0.283). Furthermore, this trend changed after 10 layers were sputtered (measured at a depth of ca. 130 nm) when the DLC-1 displayed a lower sp^2/sp^3 ratio (0.201), although the DLC-2 had a higher sp^2/sp^3 ratio (0.343). This indicated that the inner surface of the DLC-2 had a more diamond-rich structure than the DLC-1. Nanoindentation testing suggested that the DLC-1 had better mechanical properties than the DLC-2 at the external surface region, while the DLC-2 seemed to have better mechanical properties than the DLC-1 at the inner surface region. These findings are supported by the sp^2/sp^3 ratios measured at the different depths in this study, because the diamond structure is harder than the graphite structure [16]. Quantitative analysis of hydrogen in a DLC can be performed by elastic recoil measurements [27]. Using this technique, the average hydrogen content of DLC-2 (27%) was slightly higher than that of DLC-1 (23%). A higher hydrogen content of a DLC coating layer can lead to a higher hardness and elastic modulus [28,29], which may influence wear rate and frictional properties.

Most DLC films are harder than metallic materials. DLC coatings using PBIID methods provide hardnesses ranging from 6 to 20 GPa, depending on the deposition conditions [16,18,19]. The hardness of the DLC layers determined by nanoindentation testing in this study ranged from 9.18 to 9.69 GPa (when measured at a depth of ca. 70 nm), which is much higher than the 6.4 GPa measured by nanoindentation testing under the 20-mN load of the as-received stainless steel orthodontic wire. Additionally, the DLC layers showed a much higher elastic modulus compared with non-coated stainless steel orthodontic wires [30], which should influence the elastic modulus of whole archwires. This is supported by the three-point bending results of the present study. The DLC-coated wire exhibited a significantly higher elastic modulus (by 6%–11% as measured by the three-point bending test) than the non-coated wire. Fortunately, variation of this level may not influence clinical orthodontic tooth movement because a wide range of initial orthodontic forces (18–1500 gf) has been proposed as the optimum force for orthodontic tooth movement, and evidence is lacking regarding the optimal force level [31]. Three-point bending at a span of 1.0 mm caused the coating layer to detach from the inner core for only the DLC-1 wire. None of the coatings of the DLC-2 wires were damaged, probably because the DLC-2 coating had better mechanical properties and adhesion.

Several recent studies of DLC coating reported excellent frictional properties [9–12,18–20], fine cell growth with non-cytotoxicity [18], less bacterial adhesion [32] and inhibited biofilm formation on the metal with DLC coatings [33]. Similarly, the progressive-load scratch test in the present study revealed that both DLC-coated disk specimens (DLC-1, DLC-2) displayed significantly lower frictional forces than the non-coated disk specimens. One explanation for this behavior is that the DLC layer, with higher hardness due to the diamond-rich structure, produced lower frictional forces because of a lower wear rate [16]. Additionally, the hydrogen content might have contributed to lower friction under the dry condition because of the elimination of free σ-bonds on the surface [12]. However, only DLC-2 produced significantly lower frictional force than the non-coated case in the drawing-friction test with a 10° positioning of the bracket under the wet condition. This was attributed to partial rupture of the coating of DLC-1, causing increasing wire-binding at the edge of the bracket [34], thereby increasing the frictional force. Crack initiation and ruptured coating regions were not observed for DLC-2, which suggested that the DLC-2 coating had good flexibility as a functionally-graded material with outstanding adhesion to the orthodontic stainless steel substrate. Additionally, the hydrogen content of the DLC layers might be important under the wet condition. Water molecules might react with a hydrogenated DLC coating to form oxygen-containing hydrophilic groups on the surface that could provide lubrication for the sliding counter surface [21,22]. Another possibility is that hydrogen-terminated surfaces of a hydrogenated DLC coating may interact through weak van der Waals forces [16,22].

The improved frictional properties demonstrated in this work for the DLC-coated samples suggest that tooth movement by sliding mechanics using DLC-coated stainless steel wire may be superior to that using conventional stainless steel wire. However, further randomized controlled trials are required to assess the clinical efficacy.

5. Conclusions

Two types of DLC coatings (DLC-1, DLC-2), differing in hydrogen content, sp^2/sp^3 ratio and mechanical properties, were deposited on orthodontic stainless steel substrates. These coatings affected in vitro bending and frictional properties. DLC-2 showed superior frictional properties, good flexibility and adhesion to the stainless steel. A DLC coating with a higher hydrogen content may provide a better orthodontic wire.

Author Contributions: M.I. conceived of and designed the experiments. T.M., M.I. and M.K. performed the experiments. M.I., T.M. and I.M. wrote the paper.

Funding: This study was partially supported by a Grant-in-Aid Scientific Research from the Ministry of Education, Culture, Sports, Science and Technology, Japan (No. 18K09864).

Acknowledgments: The authors thank Masahiko Sugihara and Yoshimi Nishimura at Kurita Seisakusho for their expert technical assistance with the DLC coating procedure.

Conflicts of Interest: The authors declare no conflict of interest.

References

1. Iijima, M.; Zinelis, S.; Papageorgiou, S.N.; Brantley, W.; Eliades, T. Orthodontic brackets. In *Orthodontic Application of Biomaterials*; Eliades, T., Brantley, W., Eds.; Woodhead Publishing: Sawston, UK, 2017; pp. 75–96.
2. Iijima, M.; Endo, K.; Yuasa, T.; Ohno, H.; Hayashi, K.; Kakizaki, M.; Mizoguchi, I. Galvanic corrosion behavior of orthodontic archwire alloys coupled to bracket alloys. *Angle Orthod.* **2006**, *76*, 705–711. [PubMed]
3. Bakhtari, A.; Bradley, T.G.; Lobb, W.K.; Berzins, D.W. Galvanic corrosion between various combinations of orthodontic brackets and archwires. *Am. J. Orthod. Dentofac. Orthop.* **2011**, *140*, 25–31. [CrossRef] [PubMed]
4. Genelhu, M.C.; Marigo, M.; Alves-Oliveira, L.F.; Malaquias, L.C.; Gomez, R.S. Characterization of nickel-induced allergic contact stomatitis associated with fixed orthodontic appliances. *Am. J. Orthod. Dentofac. Orthop.* **2005**, *28*, 378–381. [CrossRef] [PubMed]
5. Amini, F.; Jafari, A.; Amini, P.; Sepasi, S. Metal ion release from fixed orthodontic appliances—An in vivo study. *Eur. J. Orthod.* **2012**, *34*, 126–130. [CrossRef] [PubMed]

6. Sifakakis, I.; Eliades, T. Adverse reactions to orthodontic materials. *Aust. Dent. J.* **2017**, *62*, 20–28. [CrossRef] [PubMed]
7. Kusy, R.P.; Tobin, E.J.; Whitley, J.Q.; Sioshansi, P. Frictional coefficients of ion-implanted alumina against ion-implanted beta-titanium in the low load, low velocity, single pass regime. *Dent. Mater.* **1992**, *8*, 167–172. [CrossRef]
8. Burrow, S.J. Friction and resistance to sliding in orthodontics: A critical review. *Am. J. Orthod. Dentofac. Orthop.* **2009**, *135*, 442–447. [CrossRef] [PubMed]
9. Muguruma, T.; Iijima, M.; Brantley, W.A.; Mizoguchi, I. Effects of a diamond-like carbon coating on the frictional properties of orthodontic wires. *Angle Orthod.* **2011**, *81*, 141–148. [CrossRef] [PubMed]
10. Muguruma, T.; Iijima, M.; Brantley, W.A.; Nakagaki, S.; Endo, K.; Mizoguchi, I. Frictional and mechanical properties of diamond-like carbon-coated orthodontic brackets. *Eur. J. Orthod.* **2013**, *35*, 216–222. [CrossRef] [PubMed]
11. Akaike, S.; Hayakawa, T.; Kobayashi, D.; Aono, Y.; Hirata, A.; Hiratsuka, M.; Nakamura, Y. Reduction in static friction by deposition of a homogeneous diamond-like carbon (DLC) coating on orthodontic brackets. *Dent. Mater. J.* **2015**, *34*, 888–895. [CrossRef] [PubMed]
12. Zhang, H.; Guo, S.; Wang, D.; Zhou, T.; Wang, L.; Ma, J. Effects of nanostructured, diamondlike, carbon coating and nitrocarburizing on the frictional properties and biocompatibility of orthodontic stainless steel wires. *Angle Orthod.* **2016**, *86*, 782–788. [CrossRef] [PubMed]
13. Iijima, M.; Yuasa, T.; Endo, K.; Muguruma, T.; Ohno, H.; Mizoguchi, I. Corrosion behavior of ion implanted nickel-titanium orthodontic wire in fluoride mouth rinse solutions. *Dent. Mater. J.* **2010**, *29*, 53–58. [CrossRef] [PubMed]
14. D'Antò, V.; Rongo, R.; Ametrano, G.; Spagnuolo, G.; Manzo, P.; Martina, R.; Paduano, S.; Valletta, R. Evaluation of surface roughness of orthodontic wires by means of atomic force microscopy. *Angle Orthod.* **2012**, *82*, 922–928. [CrossRef] [PubMed]
15. Kawaguchi, K.; Iijima, M.; Endo, K.; Mizoguchi, I. Electrophoretic deposition as a new bioactive glass coating process for orthodontic stainless steel. *Coatings* **2017**, *7*, 199. [CrossRef]
16. Fontaine, J.; Donnet, C.; Erdemir, A. Fundamentals of the tribology of DLC coating. In *Tribology of Diamond-Like-Carbon Films*; Donnet, C., Erdemir, A., Eds.; Springer: New York, NY, USA, 2008; pp. 139–154.
17. Roy, R.K.; Lee, K.R. Biomedical applications of diamond-like carbon coatings: A review. *J. Biomed. Mater. Res. B Appl. Biomater.* **2007**, *83*, 72–84. [CrossRef] [PubMed]
18. Kobayashi, S.; Ohgoe, Y.; Ozeki, K.; Hirakuri, K.; Aoki, H. Dissolution effect and cytotoxicity of diamond-like-carbon coating on orthodontic archwires. *J. Mater. Sci. Mater. Med.* **2007**, *18*, 2263–2268. [CrossRef] [PubMed]
19. Bentahar, Z.; Barouins, M.; Clin, M.; Bouhammar, N.; Boussirik, K. Tribological performance of DLC-coated stainless steel, TMA and Cu-NiTi. *Int. Orthod.* **2008**, *6*, 335–342. [CrossRef]
20. Huang, S.Y.; Huang, J.J.; Kang, T.; Diao, D.F.; Duan, Y.Z. Coating NiTi archwires with diamond-like carbon films: Reducing fluoride-induced corrosion and improving frictional properties. *J. Mater. Sci. Mater. Med.* **2013**, *24*, 2287–2292. [CrossRef] [PubMed]
21. Okubo, H.; Tsuboi, R.; Sasaki, S. Frictional properties of DLC films in low-pressure hydrogen conditions. *Wear* **2015**, *340*, 2–8. [CrossRef]
22. Zhang, T.F.; Xie, D.; Huang, N.; Leng, Y. The effect of hydrogen on the tribological behavior of diamond like carbon (DLC) coating sliding against Al_2O_3 in water environment. *Surf. Coat. Technol.* **2017**, *320*, 619–623. [CrossRef]
23. *ISO 14577-1:2002 Metallic Materials–Instrumented Indentation Test for Hardness and Materials Parameters–Part 1: Test Method*; International Organization for Standardization: Geneva, Switzerland, 2002.
24. Antunes, R.A.; de Lima, N.B.; Rizzutto Mde, A.; Higa, O.Z.; Saiki, M.; Costa, I. Surface interactions of a W-DLC-coated biomedical AISI 316L stainless steel in physiological solution. *J. Mater. Sci. Mater. Med.* **2013**, *24*, 863–876. [CrossRef] [PubMed]
25. Tai, F.C.; Lee, S.C.; Wei, C.H.; Tyan, S.L. Correlation between I_D/I_G ratio from visible raman spectra and sp^2/sp^3 ratio from XPS spectra of annealed hydrogenated DLC film. *Mater. Trans.* **2006**, *47*, 1847–1852. [CrossRef]

26. Ahmed, M.H.; Byrne, J.A.; McLaughlin, J.A.D.; Elhissi, A.; Ahmed, W. Comparison between FTIR and XPS characterization of amino acid glycine adsorption onto diamond-like carbon (DLC) and silicone doped DLC. *Appl. Surf. Sci.* **2013**, *273*, 507–514. [CrossRef]
27. Torrisi, L.; Cutroneo, M. Elastic recoil detection analysis (ERDA) in hydrogenated samples for TNSA laser irradiation. *Surf. Interface Anal.* **2016**, *48*, 10–16. [CrossRef]
28. Ronkainen, H.; Varjus, S.; Koskinen, J.; Holmberg, K. Differentiating the tribological performance of hydrogenated and hydrogen-free DLC coatings. *Wear* **2001**, *249*, 260–266. [CrossRef]
29. Papakonstantinou, P.; Zhao, J.F.; Lemoine, P.; McAdams, E.T.; McLaughlin, J.A. The effects of Si incorporation on the electrochemical and nanomechanical properties of DLC thin films. *Diam. Relat. Mater.* **2002**, *11*, 1074–1080. [CrossRef]
30. Iijima, M.; Muguruma, T.; Brantley, W.A.; Mizoguchi, I. Comparison of nanoindentation, 3-point bending, and tension tests for orthodontic wires. *Am. J. Orthod. Dentofac. Orthop.* **2011**, *140*, 65–71. [CrossRef] [PubMed]
31. Ren, Y.; Maltha, J.C.; Kuijpers-Jagtman, A.M. Optimum force magnitude for orthodontic tooth movement: A systematic literature review. *Angle Orthod.* **2003**, *73*, 86–92. [PubMed]
32. Muguruma, T.; Iijima, M.; Nagano-Takebe, F.; Endo, K.; Mizoguchi, I. Frictional properties and characterization of a diamond-like carbon coating formed on orthodontic stainless steel. *J. Biomater. Tissue Eng.* **2017**, *7*, 119–126. [CrossRef]
33. Cazalini, E.M.; Miyakawa, W.; Teodoro, G.R.; Sobrinho, A.S.S.; Matieli, J.E.; Massi, M.; Koga-Ito, C.Y. Antimicrobial and anti-biofilm properties of polypropylene meshes coated with metal-containing DLC thin films. *J. Mater. Sci. Mater. Med.* **2017**, *28*, 97. [CrossRef] [PubMed]
34. Muguruma, T.; Iijima, M.; Yuasa, T.; Kawaguchi, K.; Mizoguchi, I. Characterization of the coatings covering esthetic orthodontic archwires and their influence on the bending and frictional properties. *Angle Orthod.* **2017**, *87*, 610–617. [CrossRef] [PubMed]

MDPI
St. Alban-Anlage 66
4052 Basel
Switzerland
Tel. +41 61 683 77 34
Fax +41 61 302 89 18
www.mdpi.com

Coatings Editorial Office
E-mail: coatings@mdpi.com
www.mdpi.com/journal/coatings